UNIDO
P. Koenz
Director
Division of Policy Coordination

M.C. Verghese
Head, Chemical Industries Section
Industrial Operations Division

H. May
Senior Industrial Development
  Officer
Chemical Industries Section, IOD

M. El-Halfawy
Interregional Adviser
Chemical, Petrochemical and
  Fertilizers, IOD

C.E. Keleti
Senior Industrial Development
  Officer

M. Maung
Industrial Development Officer
Chemical Industries Section, IOD

M.S. Al-Ali
IBRD/UNIDO Investment Cooperative
  Programme Office, IOD

List of Observers from
  Inter-Governmental
  Organizations

ARPEL
Carlos Vanrell Pastor
Secretary General

IEA
Christopher Laidlow
Division for Producer/
  Consumer Relations

James W. Reddington
Chief Economist
OECD

OAPEC
Ahmed Nour El-Din
Director of Petroleum
  Industrialization Dept.

Khalid Adham
Economic Research

OPEC
Ali Jaidah
Secretary General

Adnan Al-Janabi
Acting Chief,
  Economics Dept.

Abderrezzak Ferroukhi
Senior Economist
Economics Dept.

Winston F. Dublin-Green
Senior Exploration Officer
Technical Dept.

Kanayo N. Ozoemene
Senior Information Officer
Information Dept.

Djemal Berrouka
Head Office of the
  Secretary General

Secretariat of the
  Symposium

Vladimir Baum
Representative of the
  Secretary-General of the
  United Nations and Director
  of the Symposium
Director, Centre for Natural
Resources, Energy and
  Transport

Howard C. Brand
Assistant Director, Energy
  and Mineral Development
  Branch, CNRET

Nasuh Adib
Chief, Energy Section, CNRET

Charalambos Constantinou
Secretary and General
  Rapporteur of the
  Symposium

Julieta Halley
Administrative Officer

F.W. Herold
Conference Co-ordinator
UNIDO

Errata for UNCNRET/STATE PETROLEUM ENTERPRISES IN
DEVELOPING COUNTRIES

pp. vi, 92, 210: Piet Harvono should be Piet Haryono
p. 5, line 7: 50 percent should be 60 percent
p. 94, line 4: natural should be national
p. 153, line 20: add "Case A"
p. 174, line 31: 25 percent should be 250 percent
p. 194: delete "KHMER REPUBLIC/Petronas Oil and Gas Agency"
p. 196: delete "TAIWAN/Chinese Petroleum Corporation, Taipei
p. 210: delete entries for Ervin Laszlo and Joel Kurtzman
p. 211, line 1: Irani should be Iraqi

LIST OF PARTICIPANTS

Afghanistan
Ansar Skandary
Director of Planning Dept.
Afghan National Petroleum Co.

Algeria
Farouk Bengalouze
Sous-directeur économique et
    financier, Ministère de
    l'Energie et des industries
    Pétrochimiques

Argentina
Raul Ondarts
Presidente
Yacimientos Petrolíferos Fiscales
    Sociedad del Estado

Benin
Edmond-Pierre Amoussou
Directeur Général de la SONACOP
Société Nationale de
    Commercialisation des Produits
    Petroliers

Eusèbe Sossouhounto
Sécretaire Général de la SONACOP

Maurice Tauziet
Chef du Service des Hydrocarbures
Office Beninois des Mines

Botswana
Raphael M.L. Sikwane
Chief Commercial Officer
Ministry of Commerce and Industry

Brazil
H.S. Albertazzi Drummond
Head, Budget Dept.
Petrobras

Burundi
Grégoire Kivueugu
Géologue
Ministère de la Géologie,
    Mines et Industrie

Ecuador
Luis A. Jativa
Gerente General de la
    Corporación
Estatal Petrolera Ecuatoriana
CEPE

Marcelo Herdoiza
Director de Finanzas, CEPE

Franklin Astudillo
Jefe, Departamento de Auditoria
Dirección General de
    Hidrocarburos
Ministerio de Recursos Naturales

Galo Bonilla Santillan
Jefe, Departamento de
    Exploración
Dirección General de
    Hidrocarburos
Ministerio de Recursos Naturales

El Salvador
C.H. Henríquez Lopez
Consejero Jurídico
Comisión Nacional de Petroleo

Vladimiro Villalta
Secretario Ejecutivo
Comisión Nacional de Petróleo

Ethiopia
Kebede Akalewold
General Manager
Ethiopian Petroleum Corporation

Fiji
Peter Johnston
Senior Planning Officer
Ministry of Economic Planning

Gabon
N'Dao-Rilogue
Ingénieur au Ministère des
    Mines et de L'Energie

State Petroleum
Enterprises in
Developing Countries

Pergamon Policy Studies on U.S. and International Business

Related Titles

PERGAMON
POLICY
STUDIES

ON U.S. AND
INTERNATIONAL BUSINESS

State Petroleum Enterprises in Developing Countries

United Nations Centre for Natural Resources, Energy and Transport

Published for the United Nations

Pergamon Press
NEW YORK • OXFORD • TORONTO • SYDNEY • FRANKFURT • PARIS

338.27282
U586

Pergamon Press Offices:

U.S.A   Pergamon Press Inc., Maxwell House, Fairview Park, Elmsford, New York 10523, U.S.A.

U.K.   Pergamon Press Ltd., Headington Hill Hall, Oxford OX3 0BW, England

CANADA   Pergamon of Canada Ltd., 150 Consumers Road, Willowdale, Ontario M2J 1P9, Canada

AUSTRALIA   Pergamon Press (Aust) Pty. Ltd., P.O. Box 544, Potts Point, NSW 2011, Australia

FRANCE   Pergamon Press SARL, 24 rue des Ecoles, 75240 Paris, Cedex 05, France

FEDERAL REPUBLIC   Pergamon Press GmbH, 6242 Kronberg/Taunus,
OF GERMANY   Pferdstrasse 1, Federal Republic of Germany

Library of Congress Cataloging in Publication Data

United Nations Symposium on State Petroleum Enterprises
in Developing Countries, Vienna, 1978.
State petroleum enterprises in developing countries.

(Pergamon policy studies)
Bibliography: p.
Includes index.
1. Underdeveloped areas—Petroleum industry and
trade—Government ownership—Congresses. I. United
Nations. Centre for Natural Resources, Energy, and
Transport. Dept. of Technical Co-operation for
Development. II. Title.
HD9560.1.U54 1978   338.2'7'282091724   79-20681
ISBN 0-08-025126-9

Printed in the United States of America

11331

Contents

CONTENTS   vii

NA

Foreword

The number, scope, and importance of state petroleum enterprises is growing rapidly as energy problems confront governments in all groups of countries.

The papers in this volume were presented at the United Nations Symposium on State Petroleum Enterprises in Developing Countries which was hosted by the Government of Austria at the Congress Centre, Hofburg, Vienna March 7-15, 1978. The Symposium was an important initiative of the Centre for Natural Resources, Energy, and Transport (CNRET) of the United Nations Secretariat. In fact, it was the first international meeting on this important topic. This initiative was taken on the basis of experience gained in execution of relevant projects in developing countries and in the organization of similar meetings in related subjects. These meetings included the United Nations Interregional Seminar on Petroleum Refining in Developing Countries (New Delhi, January 22-February 3, 1973), (1) and the United Nations Meeting on Co-operation Among Developing Countries in Petroleum (Geneva, November 10-20, 1975), (2) which also demonstrated the increasing involvement of state petroleum enterprises in some of the developing countries. The United Nations Development Programme has provided essential financial support for all these activities.

A Symposium of this size could not have been organized without the generous financial support of the Government of Austria which was augmented by the provision of excellent conference facilities. All participants expressed their deep gratitude for the hospitality of the Austrian Government and the Austrian people. While all those involved cannot be mentioned by name, special reference must be made to the untiring efforts of Ludwig Bauer, President of OMV AG, the state petroleum enterprise of Austria. He participated actively in the planning of the Symposium during consultations with the Secretariat at United Nations Headquarters in New York and, of course, during the deliberations of the Symposium in Vienna.

Special thanks are due to all the Rapporteurs and the Keynote Speakers whose valuable contributions are included in this volume as they were submitted prior to the convening of the Symposium; but their contribution was further extended by their addresses at the Symposium. In several cases they provided the main impetus of the discussions which took place.

In meetings such as this, success utlimately depends on the involvement of the participants and the discussions which follow. The presentation of each topic at the Symposium amply demonstrated the willingness of the developing countries to share a frank exchange of information and experiences and, perhaps more importantly, to be challenged by new ideas, especially ideas relating to further cooperation among state petroleum enterprises.

Several participants submitted country papers with information and analysis of their own state petroleum enterprises, and these were supplemented with oral presentations. Lack of space in this volume has prevented the publication of these papers, but there is no doubt that a separate volume is required with comprehensive information on each of the state petroleum enterprises if deeper analysis of their performance and their future is to be undertaken.

A comprehensive summary of the discussions was included in the draft report of the Symposium which was submitted on the closing day, and a "Conference review" of the Symposium has been published in the Natural Resources Forum. (3) Only the summary conclusions as adopted by the Symposium are therefore presented below:

> The Symposium has provided, for the first time, an opportunity for the state petroleum enterprises of the developing countries to exchange experiences about their organization, management, and activities, which encompass all sectors of the petroleum industry. They were able also to examine actual and potential cooperative agreements among themselves and with other institutions.

> Drawing on the lessons learned from past experience, the Symposium has formulated forward-looking suggestions in the field of cooperation for the strengthening of state petroleum enterprises which will enable them to achieve the goals of the developing countries within the framework of a New International Economic Order.

> The Symposium has provided an impressive demonstration of the growing importance of state petroleum enterprises as effective instruments of national economic development. It has also pointed to the role they must play increasingly in the world economy.

Among the fundamental problems identified at the Symposium was the disappointing level of petroleum exploration activity in the developing world. Given the possibility of future petroleum shortages and, in view of the untapped oil potential of the developing world, it was suggested that cooperative measures be taken by the industrialized and oil producing countries, under the aegis of the United Nations, to examine the possibility of multilateral financing of petroleum exploration in the developing countries which are experiencing difficulty in attracting risk capital.

In view of the wide disparity of technical, managerial, and administrative experience among state petroleum enterprises - in both the developed and the developing world - and given the need for technology transfer and training experience by the less developed countries, it was proposed that a central clearing facility for the exchange of information be established, possibly with the support of the United Nations. In this connection, the Symposium noted the generous offer of the Austrian State Petroleum Administration to provide a temporary facility.

Special attention was paid by the Symposium to research and training by state petroleum enterprises. Actual situations varied from one country to another, but considerable progress had already been achieved and some developing countries were not only practically self-sufficient in this regard, but they were able to help other developing countries as well. It was necessary for an organization like the United Nations to undertake a survey of all research and training centers in developing countries and, if appropriate, to promote cooperation in this field among the developing countries either through the use of existing centers or the establishment of regional institutions.

The Symposium stressed the possibilities of enhancing regional cooperation among state petroleum enterprises. Among several suggestions, the statement of the majority of the African participants on the establishment of a Regional Organization for Mutual Co-operation and Assistance in Oil Matters in Africa was a notable contribution.

It was generally felt that the momentum created by the Symposium in fostering contacts among state petroleum enterprises in the developing countries should be maintained and that all means for promoting closer links among such state petroleum enterprises should be encouraged.

Finally, the contribution of Steven Hein, Associate Economic Affairs Officer with CNRET, who edited and arranged for the retyping of all the papers for the publication of this volume is appreciated.

> C. Constantinou
> Secretary and General Rapporteur
> United Nations Interregional Sympo-
>   sium on State Petroleum Enter-
>   prises in Developing Countries;
> Senior Economic Affairs Officer
> Energy Section
> Energy and Mineral Development
>   Branch
> Centre for Natural Resources,
> Energy and Transport

NOTES

(1) The United Nations Interregional Seminar on Petroleum Refining in Developing Countries, (New Delhi: Indian Oil Corporation, 1973).

(2) Petroleum Co-operation Among Developing Countries, United Nations publication, Sale No. E.77.II.A.3.

(3) C. Constantinou, "State Petroleum Enterprises in Developing Countries," Natural Resources Forum 3, no. 1 (October 1978).

Introduction

We live in rapidly changing times, and perhaps in no other sector of the world economy does this appear more clearly than in the field of energy. Often, the seeds of such change lie in the past, they evolve slowly, and suddenly we face a radical transformation. In both the industrialized and the developing countries, governments - regardless of social and economic philosophy - have long taken direct interest in energy matters despite "professions of faith" often of a very different character. However, government involvement in energy has taken new forms and intensified, particularly in the last decade or so.

In the developing world, one of the most impressive expressions of this greatly heightened government concern for and involvement in energy problems is undoubtedly the blossoming of state petroleum enterprises. Although the first state oil corporation of a developing country was founded in Argentina as early as 1922, and Mexico nationalized its oil industry in 1938, the trend toward the creation of national oil entities has received an overwhelming momentum only in the recent past. By the end of 1977, some 80 state petroleum enterprises were operating in the developing countries; nor are state petroleum enterprises new to the industrial world, though here, too, there are quite a few notable additions which are of recent vintage. Austria has had a very efficient state petroleum enterprise since the mid 1950s, and new ones have been established in Norway, Canada, the United Kingdom, and elsewhere.

A number of factors have contributed to this trend, and the articles in this volume provide additional insight into the reasons why state petroleum enterprises have been created, the aims they pursue, and the direction they are expected to follow.

One of the main motives for their establishment is the perception that oil will remain the predominant source of energy for a good number of years to come, and that adequate supply of oil will continue to be crucial for economic growth. In this connection, state petroleum enterprises become an important instrument of government energy policy. Moreover, in the determination to control their own fate, the developing countries are enforcing the principle of permanent sovereignty over natural resources as one of the basic tenets of a New International Economic Order.

In the 1970s, the application of the principle of permanent sovereignty over natural resources and, implicitly, the establishment and growth of state petroleum enterprises have brought about a drastic change in the balance of power among those who used to control the lion's share of world petroleum resources and the producing countries. This is best illustrated by the fact that from 1963 to 1975, outside the United States, Canada and the centrally planned economies, public sector control rose from 9 to 62 percent in production; from 14 to 24 percent in refining, and from 11 to 21 percent in marketing. This trend is continuing and, as uneven as it may be, it affects in different ways different countries in both the developed and developing world.

Furthermore, the technologically advanced and complex petroleum business is an important lever in acquiring technical skills required to build a modern society. Through various linkages, petroleum enterprises provide an impetus for the growth of other industries such as petrochemicals or the manufacture of equipment for petroleum operations. A few state petroleum enterprises are already using their technological capabilities in research and development of alternative or new energy sources. Finally, the more successful enterprises contribute increasingly to the treasury and, thus, to the development of the national economy as a whole.

The tasks entrusted to state petroleum enterprises vary greatly from one country to another, depending on the country's natural endowment, the level of development of its petroleum industry, the socioeconomic setting in which the enterprise operates, whether the country is an importer or an exporter of this precious commodity, the size of the domestic market, and the direction in which the economy is moving. Moreover, state petroleum enterprises will be called upon to shoulder tasks beyond those of a purely national character as they increasingly become an important link in the complex network which enables the international community to meet its energy requirements. In this regard, we perceive only vaguely, as yet, the new forms of cooperation in energy as one of the important features of new economic relationships.

What is it then that we can expect from the state petroleum enterprises in the developing world? What are the practical tasks they are called upon to face, and how are they, in their own ways, addressing the problems of the future?

In answering these questions, one cannot help but be struck by the danger of easy or superficial generalizations. While some state petroleum enterprises already have a long and distinguished history, others are only in the process of being created. Some of these corporations are among the most important producers in the world. Others are only entering the fields of marketing or refining and are, at the same time, expected to supply their countries with the minimum of basic requirements. Some of these organizations are already fully integrated and some have expanded their operations outside their borders. And again, while a number of state petroleum enterprises have reached a high level of technical and managerial expertise, others are still in a rather early learning stage. Where some have had relatively sufficient time to grow progressively - starting with marketing, information gathering, and assuming gradual control over the supply function and production - in many cases this process will have to be telescoped within a very short period of time. Yet, in spite of all these differences, there is sufficient common ground for state petroleum enterprises to warrant a fruitful exchange of experiences, opportunity to learn from one another, and to explore the possible avenues for cooperation.

Energy consumption in the developing world is dominated by petroleum in both petroleum-exporting and petroleum-importing countries. In 1975, the petroleum deficient countries as a whole depended on oil for almost two-thirds of their commercial energy requirements. This figure is an average heavily weighted by a few countries that consume large quantities of indigenous coal or natural gas or that have managed to harness important hydropower potential. The overwhelming majority of petroleum deficient countries are dependent on petroleum for at least three-quarters of their energy needs and most of it has to be imported. At present only 10 of these countries produce any oil.

Energy requirements of the developing countries will necessarily outpace the overall growth rate of their economies as commercial energy sources gradually replace the noncommercial ones. Consequently, the energy/GDP coefficient will be higher than one. According to a rather moderate forecast by the French Petroleum Institute, by the year 2000 developing countries as a group are expected to augment their energy consumption by four to six times the present level. As sanguine as one may be about the harnessing of alternative or new sources of energy, during this historically short period of some twenty years petroleum will have to provide the largest share of energy requirements.

These days, rather gloomy prognoses about the future availability of oil are in fashion. So far, it is unclear whether there is sufficient solid evidence either to accept or dismiss them out of hand. It is, nonetheless, certain that a sustained global exploration effort is imperative. Geologists affirm that less than half the ultimate recoverable hydrocarbon resources have been discovered to date. Perhaps as much as two-thirds of the global resources which remain to be found are to be located in the developing countries. About half of these prospective discoveries will probably be made offshore in areas where relatively little exploration has been carried out.

Statistics throw an interesting light on exploration trends around the world. While more finds are to be expected in the developing than in the developed countries, the exploration activities in the Third World have conspicuously lagged behind those in the industrialized countries. In 1975, outside the centrally planned economies, 60 percent of all seismic party-months took place in the industrialized countries (52 percent in North America alone), 24 percent in oil exporting countries, 13 percent in the 10 oil producing but still importing developing countries, and only 3 percent in 80 nonproducing developing countries. This picture is even more telling when one considers the number of exploratory wells drilled. More than nine-tenths have been sunk in the industrial countries (86 percent in North America) and less than one per cent in those developing countries where no oil has been produced as yet. It is estimated that, in 1976, no more than a mere $200 million was spent on exploratory drilling in these 80 countries. Furthermore, it is a matter of serious concern that a decrease in the share and in absolute numbers of wells drilled in the developing countries has been registered over the last couple of years. However, these figures should not obscure the fact that, though only in relatively few countries, impressive national efforts have brought about very gratifying results. But, on the whole, it appears clear that state petroleum enterprises, governments, and the international community confront a huge task in exploration to which they must address their fullest attention.

Obviously, the twin questions of technology and finance are key issues in promoting the exploration effort. They are even more important in production and downstream activities. Ideally, phased exploration programs should not present insurmountable financing difficulties, as they can be initiated with relatively modest funding, although, by the very nature of these ventures, one cannot avoid an element of risk or gamble. Through diversification and integration, the international oil industry has managed to minimize (or average out) this risk element. Some successful state petroleum enterprises in the petroleum producing countries have also been able to embark on this route. For most of the petroleum importing developing countries this is an unlikely option. The progress of modern technology can, however, help to reduce the risk element.

The acquisition and management of technology is one of the fundamental issues before state petroleum enterprises. Reading the papers included in this volume, one is impressed by the achievements of many developing countries in this regard. Nevertheless, we all know that the industrial world retains an overwhelming lead in technology. An important question is precisely that of the ways and means for state petroleum enterprises in the developing countries to gain access to advanced technology.

The growth of the international oil industry has been made possible by the use of sophisticated technologies which have required a very substantial research and development effort - to a considerable extent financed by the oil companies. However, the oil companies have by no means a monopoly over these techniques. In fact, they rely heavily on specialist service firms, contractors, construction companies, equipment manufacturers, and so on. The oil companies' strength has been in managing the various inputs, in orchestrating, under their direction, the work on the specialized parapetroleum industries whose services are available, in principle, to any client.

A number of state petroleum enterprises in developing countries have already made decisive strides in the process of mastering this complex art of acquiring and managing technology. But, generally, they are only at the beginning of a long and arduous, and I would add expensive, march. The approaches adopted in this process are as varied as the problems faced. They will range from a variety of service and management contracts to the establishment of local technical capabilities either on a national or on a regional basis, such as the OAPEC joint service companies. This process is being supported by the establishment of numerous national institutions for training and research. However, the experience which is needed for proper application of technology cannot be easily and readily handed over as a simple package to any comer. Progress will largely depend on the quality of human resources the country will be able to mobilize for this purpose and on the incentives it will provide for the necessary effort.

The papers in this volume review the approaches which have been adopted for the acquisition and adaptation of technology, the positive results achieved, and the difficulties encountered. It should help to point out the areas where cooperation among state petroleum enterprises in developing countries - or, for that matter, cooperation among developed and developing countries - could prove to be most productive. However, it can already be stated without any reservation that in those countries where state petroleum enterprises have been successful in mastering technology, they have become the nurseries which have provided skills and talent for the national petroleum industry, and they have turned into the engines of technological progress for the entire national economy.

This introduction has touched upon only two of the many problems state petroleum enterprises have to confront. State petroleum enterprises will have to deal with other very serious problems as well, whether in deciding where and what type of crude oil to buy and under what conditions, or what kind and size of refinery to build, or how to expand marketing and distribution networks, or what to do with associated gas, and so on. How well state petroleum enterprises will be able to perform such tasks will depend to a large extent on how they are organized and managed, on their relations with the country's policy-makers, on how well they are informed of their country's needs and constraints, and on the energy and petroleum situation abroad.

Problems of organization, management, and information may have to be solved even before those of financing and technology. As much as the conditions may vary in the countries where state petroleum enterprises operate, a great deal of valuable knowledge can be gleaned from the success stories as well as from the difficulties encountered or even failures.

From the papers presented here, clearer ideas have emerged on how managers of state petroleum enterprises can better assist one another in their current tasks, as well as in the formulation of long-range policies. They have led to a better perception of mutually supportive, cooperative arrangements which might be desirable and feasible. In this respect, important lessons can be learned from such organizations as the Association for Mutual Co-operation among State Petroleum Enterprises in Latin America (ARPEL), the Organization of Arab Petroleum Exporting Countries (OAPEC), and the Committee for Petroleum of the South East Asian Countries (ASCOPE). And, that there is scope in widening regional cooperation to encompass state oil enterprises from different continents is demonstrated by the statement from the African nations.

State petroleum enterprises are definitely one of the cornerstones of the world petroleum and energy economy. The symposium recorded in this volume was instrumental in defining the problems they face, the role they play today, and what additional functions they will have to assume in the future.

Vladimir Baum
Director
United Nations Inter-regional
    Symposium on State Petroleum
    Enterprises in Developing
    Countries
Director
Centre for Natural Resources,
    Energy and Transport
Department of Technical Co-op-
    eration for Development
United Nations

State Petroleum
Enterprises in
Developing Countries

I

Rationale, Development, and Organization of State Petroleum Enterprises

1 The Rationale of National Oil Companies
Paul H. Frankel

The structure of an industry and its component parts depends on the economic, social, and political circumstances prevailing at any given time and place. Thus, the oil industry's history - and its present shape - has been determined by the general conditions as they have developed during this century and as they presently exist. Therefore, the comparatively recent phenomenon of state enterprises has to be seen against the background of the previous period's setup.

The high-risk and high-investment features characteristic of all phases of the oil industry - exploration, production, transportation, refining, and marketing - have always involved a substantial degree of horizontal concentration and vertical integration. The term concentration relates to the size of the market share of the biggest operators. Although the oil industry is not as concentrated as are, for example, the automobile and aluminum industries, the market share of a few big oil enterprises has always been very large and their influence considerable.

The main reason for the predominance of these large enterprises lies in the fact that they are affected less than are smaller and localized enterprises by the potentially fatal impact of failure to find oil or unfavorable conditions in refining, transportation, or marketing phases. The reason lies in their ability to average out results from different oil fields or marketing areas, some of which are always better than others.

To be so diversified presupposes a large size, a level that usually was achieved by taking over a number of smaller firms. It is a case of survival of the fittest. Well managed, large-scale enterprises had a greater life expectancy than smaller miscellaneous firms.

Some of the features that led to horizontal concentration also called into being vertical integration. These included control of more than one phase of the industry by the same corporate or managerial entity. The motivation for such integration is twofold. A refiner that is also a crude oil producer is sure to have a captive buyer. By selling its products to a large number of ultimate customers, a refiner knows that it will, in most circumstances, be able to run its plant regularly and use

3

its capacity to a satisfactory degree. This is a characteristic known as forward integration, and is particularly useful at times when there is ample crude oil production and refining capacity available. On the other hand, at times of crisis and shortage, the marketer has greater assurance of supply if he controls a refinery, and the refiner if he has crude oil of his own (backward integration). Another advantage of integration is the opportunity to average out the usually different profitabilities of various phases. The stability and resistance to shock of a company is greater if it is not dependent solely on the earnings of any one phase but encompasses them all.

These are the real reasons for the development of the very large oil companies. It is true that in countries whose system favored competition there were always smaller and nonintegrated companies whose existence helped to loosen the oligopolistic grip of the large companies. But the activities of the so-called independents were, and remain, marginal. They are a supplement to big operators, not a substitute for them. After a while, many of the independents fade out, although they are usually replaced by newcomers or are merged into the prevailing larger networks.

Although diversification, concentration, and integration originally took place in any one country, it was only logical that they should extend into the international sphere. The transnational character of the oil industry was accentuated by the fact that the main oil-producing countries and areas (Venezuela, the Middle East, north and west Africa) had only small oil demand themselves whereas, until recently, Western Europe had hardly any oil of its own. Now the United States is also a substantial importer of oil and natural gas. The fact that an enterprise operated in more than one country but could draw supplies of crude oil from several sources, could transport them in its own tankers and pipelines, refine them at the most appropriate locations, and could rely on a whole series of markets gave the international operator a degree of flexibility and effectiveness that was hard to match.

American oil firms were the first to reach out from their home base into the world because, until the period between the two world wars, the United States was the main supplier of oil internationally, and American oil companies established themselves abroad as sales organizations handling their own oil. The British, who had little oil either in the United Kingdom or in what was first their Empire and later the Commonwealth, and who, at the beginning of this century at the height of their imperial power, were reluctant to remain dependent on the United States for oil supplies. They looked for oil in the Middle East which conveniently, before and especially after World War I, happened to be largely in a British sphere of influence.

The development of oil resources in the Middle East (and early on in Venezuela) required a good deal of initial investment, not only in the actual field work, but in the infrastructure as well, Equally important was the fact that only companies which had a substantial share in the markets of the oil - importing countries by way of integration could envisage the disposal of the large amounts of oil in these new areas whose potential could be realized only by massive and sustained exports.

Apart from the economic conditions, which called for operations to be organized by large corporate units, it was, of course, no coincidence that the seven predominant companies had the United States and Britain as their respective home bases. These were Standard Oil of New Jersey (now Exxon or Esso), Standard Oil of California (Chevron), Socony-Vacuum (Mobil), Texaco, Gulf, Anglo-Persian (BP), and Royal Dutch/Shell (50 percent Dutch, 40 percent British) and in a different way that of Compagnie Francaise des Petroles in parts of Africa was undoubtedly facilitated by the economic strength of their home countries, but it was primarily an emanation of their political, military, and naval power.

The fact that these internationally diversified companies, almost by definition, could not be expected to take fully into account the specific interests of every country in which they operated, generated the perfectly rational desire of national authorities to have direct influence in respect to energy affairs which, as time went on, became more and more important. Once governments became fully conscious of their sovereign status, they considered a situation unacceptable in which virtually all relevant decisions on disposal, procurement, and investment were made from foreign centers by remote control.

The internationally diversified enterprises - looking for oil where they could, refining and marketing where it was most economical - actually represented a system of global optimization which might have been perfectly justified in itself, yet might conceivably have been deleterious to some individual country or other. No one will accept being optimized out of the system. Thus, inevitably, evolves the concept of national optimization, whose terms of reference are primarily determined by the respective country's position and predilections.

One way to make their influence felt lies in the sphere of regulation by way of customs tariffs, import and export licenses, fiscal regimes, and so on. Such a procedure can be effective (it is the one traditionally applied in the United States), but elsewhere it was felt, in several instances, that such a purely normative system was not sufficient to assure the safeguard and furtherance of the national interest, and that direct action within the industrial sphere was required. This is the approach which almost inevitably leads to the establishment of state petroleum enterprises.

One of the motives for governments to create their own industrial instrument is their need to obtain first-hand knowledge of the working of an important industry. It was actually a Frenchman who stated that there was, in the circumstances, a need for an entreprise temoin, a company witnessing what actually happened and why.

It is obvious that the operational base of a state petroleum enterprise will be narrower than that of a transnational corporation, and that the former lacks the advantages of the latter that were described above. Thus, the continuity of the narrower-based enterprise can be assured only by its being backed by the governmental authorities that brought it into being. This is wherein lies the difference between the traditional independent and what we call a national or state enterprise.

It must be realized, however, that the term backing does not simply mean the handing out of subsidies. State petroleum enterprises can and should be profit-oriented. Their specific character is, however, fundamentally determined by two features. They are available to their government as an instrument of national policy in case of need, even if, at times, actions are to be taken that cannot be justified on purely commercial grounds. In recognition of this posture, the state will safeguard the company's standing should the going get rough for a while.

Whereas the pursuit of nationally-oriented policies is the object of all state petroleum enterprises, there is an obvious difference between the status of such companies in oil-producing and exporting countries and those in oil-importing countries.

Obviously, a country self-sufficient in oil can more easily go its own way than one dependent on others. The first western country to nationalize all oil activities by establishing a state monopoly was Mexico, which, for a long period, produced about as much oil as it needed domestically. Therefore, unlike Venezuela, it did not need the foreign oil companies for the sake of access to world markets.

For countries dependent on close and sustained contacts with the rest of the world and which do not consider it appropriate to monopolize oil activities entirely, the problem arises of coexistence with nongovernmental operators, both domestic and foreign. The decision to be taken will depend upon the general state of the industry worldwide and, in practical terms, on the usefulness of the other enterprises to one's own country.

It is obvious that the principle of global optimization cannot be dismissed altogether since it does embody valuable cost economies. Thus the existence of a state petroleum enterprise need not exclude cooperation with traditional oil industry interests. A reasonable compromise between the two forms of optimization might prove to be advantageous for all concerned. It may well be that the very existence of a state petroleum enterprise, by reducing the concern of a government lest foreign enterprises may be deriving undue advantage from the strength of their unique position, will not impede fruitful cooperation with others, but will, in fact, make it possible.

While the prevalence of transnational oil companies was in line with the global power and influence of the United States and Great Britain, the dissolution of the British Empire and the weakening of United States influence, primarily in Latin America, have brought to the fore a considerable number of sovereign power centers. The state petroleum enterprise is the appropriate expression of a new situation. Yet, as has just been pointed out, extreme isolation is unworkable. It is likely that further links will be formed between the several state enterprises. The existence of OPEC - although it is an alliance of governments - has its repercussions in the operational sphere and its contacts with state petroleum enterprises in oil-importing countries could, and most likely will in due course, become economically and politically relevant.

Also, there are already in existence some other groups such as the Assistencia Reciproca Petrolera Estatal Latinoamericana (ARPEL). The Organization of Arab Petroleum Exporting Countries (OAPEC), with a current membership of Algeria, Bahrain, Egypt, Iraq, Kuwait, Libya, Qatar, Saudi Arabia, Syria, and the United Arab Emirates, is an example of the direction in which cooperation between oil-producing countries with state petroleum enterprises might develop. Its extensive mutual information and educational system and its endeavor to devise and sponsor joint projects whose size and significance transcend the scope and sometimes the means of individual members are examples of the necessary attempt to recreate - but in a different fashion - some features of the earlier transnational operations.

Seeing these developments in their true historical context, one realizes that there is (as always) a swing of the pendulum. The break-up of global imperial systems leads to the proliferation of a large number of individual units. Once these units have become truly established, the need for a network of relationships on a new basis makes itself felt. State petroleum enterprises, which have played a vital role in what one could call the phase of emancipation and self-determination, will be called upon to provide the base for the period of coexistence and coordination whose onset we are now witnessing.

2 The Birth and Growth of the Public Sector and State Enterprises in the Petroleum Industry

C.A. Heller

The history of petroleum has been analyzed and written about from the most diverse political, philosophical, economic, and social viewpoints. This short review concentrates on one important historical phenomenon, the development of the public sector through state enterprises in the petroleum industry, particularly in the developing countries. In the context of this paper, only those entities that are totally in the hands of the government are regarded as state petroleum enterprises, irrespective of their institutional structure.*

A mere chronology will not suffice as background for a discussion of the emergence and development of state petroleum enterprises on a global level. The history of the state petroleum enterprises should be brought under focus and understood in relation to world industry as a whole.

For many years after its discovery, petroleum attracted primarily the fiscal interest of governments, since it provided a welcome source of revenue from excise taxes on some petroleum products and from production royalties, or areal rentals in those countries which reserved the subsoil rights to the state.

Petroleum history up to now has been generally divided into two basic ages: The Age of Illumination, and the Age of Energy. (1) We have now entered a third age, which one might call the Age of Conservation. The Age of Illumination lasted from the first oil discovery until the beginning of the twentieth century. It included the infancy and adolescence of the petroleum industry. Kerosene was the main product; residual fuel oil started slowly on its way. The entire world consumption in 1900 amounted to approximately 21 million tons per year (over 400,000 barrels per day) which represents the equivalent of less than 10 minutes of world consumption today.

---

*In the context of this chapter, British Petroleum (BP) is not regarded as a state petroleum enterprise since it is not wholly state controlled.

8

The Age of Energy started with the advent of the internal combustion engine at the turn of the century and the decision by the British Admiralty, taken a few years later, to switch from coal bunkers to fuel oil. The growing demand was met, among others, from the discovery of Texas as the greatest oilpatch in the world: the famous Lucas gusher in Spindletop came in during January 1901. The petroleum industry had come of age.

The Age of Conservation began in the 1970s with control of prices and production reserved to the public sector in OPEC countries. Conservation of oil and natural gas became the worldwide order of the day. As the Shah of Iran demanded at the beginning of 1974, oil should be used primarily for transportation and as a raw material for higher valued goods. The development of alternative sources of energy was recognized as inevitable. The primacy of the private petroleum industry in the world's economy started to decline. Governments became the predominant powers in international petroleum policy, the almighty regulatory factor in national petroleum policy, and last but not least, actual petroleum entrepreneurs in most countries.

But let us return to the transition period between the start of the Age of Energy and World War I. The growing importance of petroleum for the military and economic security of the industrialized countries came to the minds of the rulers of that time. Consumption increased within ten years by 100 percent. But with production capacity outdistancing consumption, the first serious disturbance in the supply/demand equilibrium occurred.

The birth of the first state petroleum enterprise came about in Austria as a consequence of such an imbalance. In 1907 and 1908, unexpected flash production was encountered in the foothills of the Carpathian Mountains southwest of Lwow, the capital of the province of Galicia. Production culminated at 2.1 million tons per year, approximately 5 percent of world production at that time. With production in the United States, Rumania, Russia, and the Dutch East Indies also on the increase, the first oil feast in world crude production occurred.

When the interested parties in Austria were unable to decide between themselves about steps to be taken to deal with this problem of surplus production, Emperor Francis Joseph approved the construction of a topping plant in Drohobycz, a refining center near the main oilfields, with a capacity of 100,000 tons per year to be owned and operated by the Austrian government.

The first "child" in the public petroleum sector had been due to an accident; the state petroleum enterprise of today holds an important position as an integral component of the industry. Global, regional, and local events, elements, and trends contributed periodically and in a variety of combinations to the growth of state petroleum enterprises. The following are a few of the factors which had a global impact:

1. The entry of Latin America with prolific production in Mexico (1910) and Venezuela (1914).

2. The breakup of the Standard Oil Trust following the decision of the United States Supreme Court of 1911. The horizontal divestiture of the Trust created a number of independent firms in various sectors of the industry.

3. The purchase by the British Government in 1914 of the majority of the shares in the Anglo-Persian Oil Co. (forerunner of British Petroleum) which was responsible for the oil discoveries in Persia (now Iran) and the contributions of the refinery at Abadan. This acquisition followed long planning by the Royal Navy and was made in order to gain independence for its oil requirements. As Winston Churchill said in 1913, it was planned to provide the British Empire with petroleum reserves located along ocean routes which the Navy could protect. The British government, through this historical act, created the first partnership between the public and the private sectors in the petroleum industry.

4. The Russian Revolution of 1917 followed by the nationalization of the entire petroleum industry in 1918. Petroleum entities were put in charge of the various sectors of the oil and natural gas industry either as government departments or as separate enterprises.

5. The Agreement of San Remo concluded in April 1920 between France, Great Britain, and the Netherlands, and later amended to satisfy the demands of the United States for the so-called "open door policy." This heralded the entry of the Western powers (with the exclusion of Italy) into the oil bounty of the Arabian lands and enabled the American oil interests to acquire concessions in the Dutch East Indies (1928).

6. The enactment on March 30, 1928 of a basically new concept in petroleum law, which centralized control of the industry in France through the so-called "monopoly by delegation." This delegation of monopolistic rights was implemented through licensing of imports and exports of crude oil and refined products, refining construction, operation, and marketing. Refinery construction in France was made attractive to improve the security of supplies and save foreign exchange. This legislation contributed to the trend of locating refineries in the consuming countries (market-oriented refineries) rather than the producing countries (source-oriented refineries).

7. The liquidation of the Ottoman Empire by World War I, and of the European colonial system after World War II. Latin American countries, most of which had enjoyed independence for more than a century, became the forerunners in attitude and policies relating to the oil industry.

Mexico started the ball rolling. Its constitution of 1917 followed the traditional Spanish-European custom of reserving the subsoil to the state. The reaction of the foreign concession holders - American and

British - to the new constitution was vehement because of the prolific oil reserves they had discovered in Mexico. Their struggle with Mexico, in which they were actively supported by their respective governments, lasted over twenty years. It had far-reaching repercussions all over Latin America.

One country after another in Latin America established state petroleum enterprises conceived as public instruments against the foreign companies. These instruments were endowed with a variety of rights and privileges ranging from total or part monopolies to operations limited to certain sectors of the industry, often parallel to the activities of private companies.

Argentina started this regional movement in 1922. It was followed by Chile (1926), Uruguay (1931), Peru (1934), and Bolivia (1936). Mexico, the torchbearer of the Latin American movement, enacted the sequestration and nationalization of its entire petroleum industry on March 18, 1938. The entry of the new state petroleum enterprise - Petroleos Mexicanos (PEMEX) - into the industry encountered serious organizational, technical, and economic obstacles for a very long time. However, we are now witnessing the long-prepared comeback of the Mexican oil industry as one of the turning points in the history of state petroleum enterprises in developing countries. Regional developments led further to the creation of state petroleum enterprises in Colombia (1951), Brazil (1954), and Ecuador (1973). It reached its high point with the nationalization of the petroleum industry in Venezuela on January 1, 1976.

Global developments lead from Mexico to Teheran. We will attempt to analyze trends and factors which favored the development of state petroleum enterprises in the Third World and the constraints which have worked against them.

However, it might be appropriate first to comment on events in Western Europe where government involvement, particularly in key industries, increased during the interwar years and afterwards. This trend resulted in the appearance of the state petroleum enterprises in several Western, industrialized countries.

Italy created Azienda General Italiana Petroli (AGIP) as a wholly state-owned entity in 1926. The basic concept behind this enterprise was to gain some independence from the British and American oil companies which dominated petroleum imports and marketing in that country. AGIP aimed at exploration for, and refining, and marketing of hydrocarbons. However, it was the appearance of Enrico Mattei, the protagonist of the state petroleum enterprise in the free market economies of the industrial world, that brought about the formation, in 1953, of Ente Nazionale Idrocarburi (ENI) as a comprehensive instrument for the national and international petroleum policy of Italy.

France established its first state petroleum enterprise under the acronym RAP (Regie Autonome des Petroles) before World War II. The company served as one of the building blocks for ELF-ERAP (Entreprise de Recherche et d'Activites Petrolieres), the present state petroleum enterprise of France.

In line with this trend towards an increase of public sector activity in the petroleum industry, a long line of state petroleum enterprises emerged in consuming nations in Western Europe. One of the enterprises, however, deserves attention for the unique circumstances which led to its birth - the State Treaty between France, the USSR, the United Kingdom, and the United States on one side and the Austrian government on the other, concluded in 1955. It specified that all of the producing fields, exploration rights, and related assets in the Soviet zone were to be turned over to Austria inter alia under the condition that their ownership could not be assigned to foreign nationals. (2)

Of course, the establishment of state petroleum enterprises in the two large oil and gas producing countries in Western Europe - Norway and the United Kingdom - also belongs to this trend of expansion of the public sector. However, the expansion of the public sector in the petroleum industry into a global movement came about as the number of newly independent countries in Africa, Asia, the Caribbean, and the Pacific increased and as the OPEC countries and other producing countries took control of their own oil industry. Certain motives for the developments in individual countries were common to the global movement, or at least similar, irrespective of the oil situation of each country.

One leading motive, economic nationalism, which could be said to have started in Mexico, followed the global line all over the world. As the Middle East Economic Survey reported in 1968:

> the emergence of the State Oil Companies in the developing world has proceeded parallel to the growth of self-awareness, of which it is in part the expression. It is based on the widespread feeling that the control of vital natural resources, on which the nation is in some cases almost entirely dependent, is not a matter which can safely be left in the hands of foreign companies whose interests do not necessarily coincide with those of the host country and which at times may be inimical to them.

This precise formulation mirrors the spirit prevailing in the Third World, which had led to the United Nations resolutions of 1962 and 1966 on Permanent Sovereignty over Natural Resources.

In analyzing other common elements which contained the spread of the state petroleum enterprise in the Third World, there is welcome support in a study by Dr. Fariborz Ghadar, an Iranian economist. (3) Ghadar selected for his analysis of the "common" elements three test cases: Indonesia, Iran, and Saudi Arabia.

Indonesia's petroleum industry goes back to the Age of Illumination in the last two decades of the nineteenth century. That country reached independence in 1949. Pertamina was formed from predecessor-state entities in 1968.

Iran, one of the oldest independent nations of the world, helped to open the Age of Energy. The National Iranian Oil Company (NIOC) was created by a law in 1951 which nationalized the oil industry, as had happened in Mexico 13 years previously.

Saudi Arabia, independent since World War I, established Petromin in 1962. The position of Saudi Arabia as owner of the largest proved oil reserves in the world dominates the beginning of the Age of Conservation.

In spite of these background differences, Ghadar's description of the growth of these three national companies can serve for a better understanding of certain common elements.

Stage I covers the era during which the potential crude oil production could not initially be exploited by the host countries because of the lack of finance, technology, and trained personnel. The major interest of the host government was that of a tax collector.

Stage II of the development occurred after the exploration stage which eliminated the risk factor. Exploration risks were left to the oil companies. When exploration proved successful, production costs were low, and with control over prices, the profits of the oil companies were high. The growing revenues of the host countries from the operations of the oil companies helped to finance development. At this stage, the push by the governments to gain more control was on. It first took the form of establishing a governmental information gathering unit. These units often provided the seed for the state petroleum enterprises. In addition to the units in Indonesia, Iran, and Saudi Arabia, mention may be made of the recommendation of the Commission of Inquiry into the Oil Industry of Trinidad and Tobago (1963/64) of which the author was a member. The Commission requested that "...the Minister of Petroleum and Mines should be kept continuously informed (by the oil companies) of the progress of their operations" in the widest possible scope, including technical, economic, financial, and other vital data. (4)

During Stage III, conflicts developed between the host country and oil companies over national development goals. This stage brought general changes in the structure and activities of the state petroleum enterprises in the three countries. Such changes were preceded by well-planned indoctrination and training of staff and personnel.

The entry of governments into an oligopolistic industry such as oil took quite some time. In most developing nations it encountered economic and technical barriers against the expansion of the public sector which were innate to specific characteristics of petroleum operations: vertical integration, economies of scale, vast capital requirements, and high risk factors in the upstream operations. Domestic marketing, with the least constraints against entry, was the first activity undertaken by most state petroleum enterprises. NIOC entered into domestic marketing shortly after the agreement of 1954 between Iran and the consortium was signed. In Indonesia, domestic marketing was taken over in 1965 and in Saudi Arabia in 1967. Once domestic markets were assured, the acquisition of existing refineries or the construction of new facilities did not encounter serious obstacles.

After the nationalization of the oil industry, Iran decided to reach out first for a smaller inland refinery. This type of refinery served as a training ground on which all levels of personnel could acquire the know-how and intricacies of refining operations. NIOC took over actual operations of the Abadan refinery and petrochemical complex only recently.

The takeover of refining in Indonesia by the public sector followed similar lines. Pertamina and its predecessors acquired six refineries from Shell between 1962 and 1971, starting with a small plant in East Java. Two other refineries went into the public sector in 1970 and 1971.

Petromin at first acquired a 75 percent interest in a new refinery in Jiddah in 1968 before building a wholly-owned refinery in Riyadh in 1975.

Exploration for hydrocarbons, the development and conservation of discovered reserves, and production itself were left until last. Even the most experienced of the state petroleum enterprises did not venture into exploration significantly except with the cooperation of transnational oil companies. Developments outside the host countries, which led to a proliferation of oil companies and easier access to technologies, were used by the state petroleum enterprises to obtain better terms from both the traditional companies and the newcomers.

The increase of ownership and operations of tankers by large producing and consuming developing countries is also part of their drive to gain more independence from the big oil companies. The tanker fleets of Indonesia, Iran, the Organization of Arab Petroleum Exporting Countries (OAPEC), and of some other developing countries represented approximately 8 million dead weight tons (dwt) at the end of 1976, or 2 percent of the world tanker fleet of approximately 380 million dwt. These countries have taken steps to increase their fleets to 18-20 million dwt by 1980 which would raise their share in world capacity to 5 percent.

Turning to the growth of the public sector in petroleum importing developing countries, one has to note that they have faced barriers to entry in various sectors of their petroleum economy. Particularly countries with very low national income, and consequently limited petroleum consumption, have been exposed to such difficulties.

Large nations like Brazil, India, and others acquired refining and marketing facilities from transnational and local private owners and constructed additional capacity themselves. They also engaged in upstream operations in various fashions inside and outside their own territories, with notable successes.

Ownership and operation of the refining and distribution sectors in other energy-importing, developing countries are quite varied. In some countries they are wholly state controlled, while in other countries they are either left entirely to transnational oil companies or are partly owned through joint ventures.

Many of the energy-importing, developing countries with unexplored oil and gas potential may face barriers similar to those faced by the state petroleum enterprises already described. For these countries in particular, cooperation with the state petroleum enterprises of petroleum-exporting, developing countries and the industrial world as well as international organizations is called for.

The events of 1973 and 1974, as well as subsequent events, led to an acceleration of the public sector as illustrated in table 2.1. This table

TABLE 2.1. Share of Ownership in the Petroleum Industry (Outside North America and Centrally Planned Economies)
(in percent)

Production	1963	1968	1972	1975
Majors	82	78	74	30
Government	9	9	12	62
Others	9	13	15	8
Refining				
Majors	65	56	54	47
Government	14	16	17	24
Others	21	23	27	29
Marketing				
Majors	62	56	54	45
Government	11	14	15	21
Others	27	31	31	34

Majors: BP, Exxon. Gulf, Mobil, Shell, Standard of California, Texaco

Source: Geoffrey Chandler, "The Innocence of Oil Companies," Foreign Policy, (Summer 1977), p. 52.

shows that the ratio of private sector to public sector participation in upstream operations has been completely reversed in the past few years. The change in the ratios in refining and marketing was, however, less drastic.

All in all, several hundred state petroleum enterprises in about 90 countries represent the public sector in the world petroleum industry. Will this trend continue? The share of the public sector in production will increase in the not too distant future, once the arrangement between the government of Saudi Arabia and ARAMCO is finalized. Beyond this, any prediction is impossible. It must not be forgotten that in exploration the unexpected is the rule.

Developments in oil refining and distribution will depend on policy decisions by both the public and the private sectors. For example, a number of transnationals have sold their refining and marketing outlets to the public sector in certain countries, perhaps in their attempt to balance their operations in the new oil equilibrium.

The large crude oil exporters will have to decide how best to secure outlets for their production on a permanent basis. Will this be done by their transformation into integrated transnational operators; joint ventures with existing organizations; cooperation with state petroleum enterprises in consuming developing countries; or by a combination of these and other choices? These and a host of other questions remain unanswered. The future power position of the public sector in the world petroleum industry will be determined by the individual and the collective strength of state petroleum enterprises. Each state enterprise represents the sovereignty of a national state. The public sector of the Third World can grow by understanding among all its members - the future, mammoth, state-owned companies and the host of other units, down to the smallest of the state enterprises.

NOTES

(1) H.F. Williamson and A.R. Daum, The American Petroleum Industry, (Evanston, Ill.: Northwestern University Press, 1959).

(2) International Petroleum Industry, Vol. 1 (New York: International Petroleum Institute Inc., 1965).

(3) F. Ghadar, The Evolution of the OPEC Strategy, (Lexington, Mass.: Lexington Books, D.C. Heath and Co., 1977).

(4) Report of the Commission of Inquiry into the Oil Industry of Trinidad and Tobago 1963-1964, (London: Andre Deutsch, 1964).

3 Problems and Prospects of State Petroleum Enterprises in OPEC Countries

Ali Jaidah

The OPEC countries rely very heavily on the production and export of crude oil and gas and, therefore, the state petroleum enterprises in OPEC occupy a special place in the economies of its member countries. However, in the wider group of developing countries, there are state petroleum enterprises which have a longer history and, in some cases, more comprehensive experience than OPEC's own enterprises. OPEC has learned, and is still learning, a great deal from the experience of enterprises such as Mexico's Pemex, the Indian Oil Corporation, the Indian Oil and Natural Gas Commission, Petrobras of Brazil, and others. Indeed, some OPEC state petroleum enterprises are only in their formative years.

Only since the beginning of the 1970s have the majority of OPEC state petroleum enterprises been closely involved in the various aspects of the oil industry. However, despite their relatively short experience, these enterprises have been playing an increasingly important role in the international oil industry and, in many cases, have taken the leading role in managing the business of producing and exporting oil in their respective countries.

Looking at the historical growth of OPEC state petroleum enter- prises, some are older than OPEC itself, while others came into existence after the birth of OPEC in order to cater to the special circumstances of an individual member country. Algeria established its enterprise shortly after independence, and Iraq after the passing of a law in 1961. However, their real development came after the 1970s with the demand for participation on the part of the existing operating companies. The task has been well accomplished. However, this success has not been achieved without some measure of problems and difficulties. Neither complete transfer of legal ownership, nor the future growth and development ever goes as smoothly as is planned on paper.

In looking at the problems which were encountered, we have to go back to the early days of the oil industry when the concession agreements concluded between the international oil majors and host countries in effect gave the companies carte blanche to carry out their operations anywhere in the country's territory. This effectively elbowed any nascent state petroleum enterprises out of all but the most unpromising fields. The advent of the so-called "independent" oil companies improved matters slightly, but the pattern remained largely unchanged.

Indeed, although many concession agreements accorded a theoretical right for each state to receive royalty crude in kind (about 12.5 percent of the oil produced in most countries), other conditions stipulated that it should be received at the posted price. Many countries, however, found that they would stand to lose in so-called "profit sharing" arrangements if they did not receive their royalty crude in cash. Due to this arrangement, state petroleum enterprises had no access to any crude oil with which to compete in a controlled market in so-called "arm's length deals." Indeed, it can be said that marketing experiences in many cases are to be learned simply from reading about them.

The legal takeover of the oil industry from the international majors, which we have witnessed in recent years in OPEC countries, has entailed increased responsibilities and, in some cases, problems too. OPEC is now standing at a new crossroads in its history as an organization and as a group of countries. It is facing a fresh challenge to formulate a new vision which goes beyond the legal control of exploration, production, and export of crude oil.

This new vision is based on the following objectives:

1. the efficient management of the oil industry by nationals of OPEC member countries at all levels;

2. the development of an indigenous technological base, backed by domestic research institutions, capable of contributing increasingly to the needs of the oil sector in member countries;

3. the speedy transformation of the role of OPEC member countries from that of raw material exporters to manufacturers by carrying out certain downstream operations, especially with regard to refining and petrochemicals. In this way, the national oil industries should become the central pivot in the process of industrialization.

Obviously, this formulation means that the state petroleum enterprises in OPEC member countries ought to play an increasingly important role, not only in the economies of their own countries, but also in the formulation of the policies and objectives of OPEC itself.

If OPEC's state petroleum enterprises are to meet this considerable challenge, there are a number of problems which they will have to resolve. In many ways, these differ very little from those facing other state petroleum enterprises in the rest of the Third World. I will, therefore, try to identify briefly some of those problems, in the hope that this exchange of ideas may help us to go some way towards resolving them. Needless to say, the member countries of OPEC differ in terms of the nature of their economies and also in the degree of direct involvement in their indigenous oil industries.

Some state petroleum enterprises control the whole of the oil sector, others depend, to varying degrees, on the cooperation of foreign companies. However, the speedy rise to eminence of the OPEC state petroleum enterprises has created serious personnel and manpower problems in all its member countries.

At all levels there is a lack of suitably trained, experienced, and specialized personnel to cover the administrative and technical operations of the oil industry. The root of the administrative and manpower problem lies in the former policies and attitudes of foreign operating companies towards personnel recruitment coupled with the limited national human resources in certain cases. To face this serious shortage, OPEC state petroleum enterprises have been seeking various interim remedies.

Some member countries are relying heavily on foreign personnel and operating companies. In the meantime, intensive training and development programs are being carried out in all skills needed for the oil industry. This task is extremely difficult and complex, since it involves a country's whole educational system and not only the organization of a specialized institute. Most OPEC member countries endeavor to send some of their nationals abroad for higher education courses and management programs at foreign universities and institutions so that at a later stage they can be of benefit to the industry either directly, or indirectly, via the educational system.

However, some member countries are already facing two problems which might hamper their ambitions in this direction. The first and lesser problem is that of the brain drain to more advanced countries, which afflicts OPEC members in the same way it does all developing countries. The second and wider problem is related to the ambitious economic development plans of the member countries which compete with the oil industry for qualified manpower. Despite the special status of the oil industry in OPEC countries, it cannot stand in the way of general economic progress by monopolizing the limited managerial and manpower resources available.

The transition from foreign control of production in OPEC countries to increasing national control has been fairly smooth, even in cases where complete nationalization has taken place. In all OPEC countries, no serious disruptions have been observed in any of the upstream phases of the oil industry. But this does not mean that we are happy with the present state of exploration and production in our countries. There is a slight slowdown in exploration activities in OPEC member countries, despite the fact that they still hold better prospects of new discoveries than many other areas. We are particularly unhappy with the old practice of the foreign oil companies of gas flaring. Indeed, OPEC's state petroleum enterprises must make a tremendous effort to put an end to all forms of flaring of associated gas. In this respect and to varying degrees they need involvement with and cooperation from the consuming countries, both with respect to technology and the appropriate pricing of gas.

With the increasing demand on OPEC oil resources, many member countries are resorting to secondary recovery techniques and other enhanced recovery methods to maintain production levels and to maximize ultimate recovery. Some member countries are allocating a great deal of their oil revenues to investments in enhanced recovery, and here, too, the state petroleum enterprises are in need of cooperation from the consumer side.

The economies of developing countries generally suffer from dependence on the export of a few, or one single natural resource. Those countries which depend on the export of a depletable natural resource are in a particularly awkward position. They must achieve their emergence from perpetual poverty before the depletion of their natural resources. Ideally, these natural resources should become raw materials for a sound industrial economy. In this context, it is only natural that OPEC countries should look to their refining and petrochemical industries as engines of change, playing a central role in their industrialization policies.

Unfortunately, the stepping-up of refining and petrochemical capacities in OPEC member countries is facing tremendous resistance from the developed countries. This resistance, if it persists, will lead OPEC to seriously doubt the willingness of the rich nations of the world to cooperate in bridging the great divide between rich and poor.

State enterprises responsible for internal energy requirements in OPEC member countries are surprisingly facing some unexpected difficulties in their own house. For instance, due to rises in energy consumption, there are imbalances in energy supply and demand in some member countries. The short-term imbalances are probably a healthy sign of change, although they cause serious inconveniences. However, in the longer term, difficulties could arise as a result of pressures from the domestic needs for energy and nonenergy uses of oil and gas. The energy-intensive industries now being developed in our member countries, as well as the petrochemical complexes envisaged for the future, might face serious shortages in the long run if depletion of petroleum resources for export is not checked in member countries with limited reserves.

Having briefly examined some of the problems now being faced by OPEC state petroleum enterprises, it is appropriate to consider the role which state petroleum enterprises in OPEC member countries, and indeed in all developing countries, may fulfill.

The establishment of state petroleum enterprises has become very fashionable these days, and in some quarters doubt has been cast on the need for state petroleum enterprises, comparing them with superfluous status symbols like national airlines. However, in the case of OPEC countries, state petroleum enterprises were created to meet a deeply felt need. During the 1950s and most of the 1960s, the transnational oil companies dominated the oil scene in the member countries. They were, in most cases, not subject to domestic clauses and exercised the power of a state. The oil sector was completely isolated from the rest of the economy and basic decisions were taken in the international centers of these companies. The only possibility for OPEC countries to have a greater say in the utilization of their basic source of livelihood was to create national institutions which could ultimately replace the foreign oil companies in every way possible. These new institutions had to start from scratch in a hostile environment and addressed themselves to the incredible task of replacing the biggest and most complex industrial giants of modern times.

In countries like Venezuela, Iran, Iraq, and Algeria some of the basic skills were already available to create state petroleum enterprises which could go directly into the oil business. In other countries, the enterprises had to feel their way at a slower pace. The international majors treated these new arrivals with a degree of aloofness and hostility. However, several independent companies and state enterprises of consuming countries were prepared to deal directly with the OPEC state petroleum enterprises. These joint ventures and contracts of the 1960s and early 1970s helped a great deal to strengthen the hand of OPEC state petroleum enterprises and to give them the necessary confidence for the future.

Now, in the latter part of the 1970s, OPEC state petroleum enterprises are oil giants in their own right. We are entering a new phase where the state petroleum enterprises play a predominant role not only in their respective domestic oil sectors, but also in the national economy as a whole. There is no doubt now that OPEC state petroleum enterprises will become the backbone of the upstream part of the oil industry, as indeed they are already in some countries. Due to their very nature, they have tended to serve the development efforts of their countries and, in some cases, have succeeded in integrating the oil sector into their national economies. In other cases, they are involved in development efforts outside the oil sector, sharing not only in the processing and industrialization of crude oil and natural gas, but also contributing to general industrialization efforts in other sectors, due to their experience and successful institutional structure.

The new phase in oil industry development, which was referred to earlier, is characterized by the perceived need for international interdependence in the energy field as a whole and, indeed, in the industrialization process in general. The era of direct conflict with the international majors is now well behind them and a new relationship is emerging between state petroleum enterprises in OPEC countries and foreign countries. The basic feature of this new relationship is greater cooperation in several fields, especially in exploration, development, and international trade of crude oil and gas. Of course, this does not that there are no differences of opinion and areas of conflict with transnational corporations, especially with regard to downstream operations, such as refining and petrochemical development.

It is also worth noting that the cooperation which started in the 1960s between OPEC countries and state petroleum enterprises of the consuming countries is still growing in a period of increased governmental involvement in energy policies. State petroleum enterprises of the consuming countries are serving as useful instruments in direct deals with OPEC state petroleum enterprises. This is true not only insofar as state petroleum enterprises of industrialized countries are concerned, but also for enterprises in the developing countries.

It is also important to consider the relations between OPEC state petroleum enterprises in energy deficient developing countries. It must be said that this area deserves a great deal of consideration and some deep rethinking is required on both sides to evolve the best form of relationship which serves the mutual interests of oil exporting countries and oil importing developing countries.

Although they are not responsible for the main diseases of the modern international economy, inflation and unstable monetary systems, OPEC countries have contributed substantially to alleviating the balance of payments problems of other developing countries, despite the fact that some are in the capital market themselves to support their own development programs. But OPEC's relationships as developing countries should go much deeper than mutual assistance. In several cases, OPEC state petroleum enterprises have contributed to the development of the energy sectors in other developing countries, such as financing refinery constructions, ensuring flexible oil supplies, exploration agreements, and others.

In the cases where state petroleum enterprises in the oil-importing developing countries are well established, such as in India, Turkey, and Brazil, cooperation is taking on very concrete and fast evolving forms which have already led to some satisfactory results. However, the resources of OPEC state petroleum enterprises are very limited and, as noted earlier, their domestic problems are many.

There is, in my view, no readily available formula to resolve this problem, but it is hoped that state petroleum enterprises of the developing consuming countries can improve their capabilities. I am sure that they will find OPEC state petroleum enterprises ready to cooperate to their mutual satisfaction.

The OPEC state petroleum enterprises are still feeling their way to maturity and, in some cases, there is still a long way to go. However, they are embarking on increased mutual cooperation in the various fields of their activities. This cooperation is in its very early stages, but it is hoped that it will evolve speedily and point the way to the type of cooperation needed among state petroleum enterprises in developing countries, whether they are exporting, self-sufficient, or importing petroleum.

There is no reason why this cooperation should not go on to contribute to the greater economic independence of the Third World. The traditional relationship of unequal exchange with the industrialized world should give way to the new pattern of mutual cooperation now evolving among the developing countries. This cooperation and solidarity should be the cornerstone of the collective demand to bring about a New International Economic Order.

4 The Role of State Petroleum Enterprises in Developing Countries: The Case of Saudi Arabia

A.H. Taher

In order to appreciate the significance of the present and future role of state petroleum enterprises in developing countries it is imperative to understand the historical events which brought about their creation and development. The case of oil exporting developing countries is of special significance in this context, and a brief review of this historical perspective will serve to illustrate this position.

The complex international political and economic developments witnessed during the last 25 years generated an awareness of the need for state petroleum enterprises in the oil exporting developing countries. Their actual creation, in most cases, took place during the past two decades - Petromin, the state petroleum enterprise of Saudi Arabia, began operation in 1962. But it is the historical events of the past decade which have resulted in the phenomenal increase in the importance of the potential role these enterprises can play in the international petroleum and energy industires.

During the past several decades and until quite recently, the transnational oil companies maintained virtually complete control over the petroleum resources of the oil exporting developing countries. These companies functioned with exceptional freedom in all phases of the industry, including determination of investment criteria, exploration, production, refining, transportation, price setting and marketing. The only sovereign right the governments exercised was collection of tax and royalty revenues.

It is now generally accepted that the traditional policies adopted by the transnationals for the exploitation and development of petroleum resources of the developing countries served mainly the interests of these companies. Little consideration, if any, was given to the aspirations of the host producing countries for natural development. Conservation of petroleum resources did not receive a high priority. Natural gas was indiscriminately flared and wasted. Refining and petrochemical industries were concentrated in consuming countries exclusively, depriving the host producing countries of the benefits of

23

industrialization. For decades, oil prices were maintained below their competitive levels. This was not only detrimental to the host producing countries' economic interests, but it also encouraged wasteful uses of depletable resources and generated unacceptable trends in the growth of the energy mix. It is a fact that during this period sovereign governments had no real control over the level of production at which their depletable petroleum wealth was produced, the prices at which their oil was sold, or to whom it was sold.

The establishment, in 1960, of the Organization of Petroleum Exporting Countries (OPEC) was in direct response to the price policies of the transnationals and was intended to protect the legitimate interests of the petroleum exporting developing countries. Intricate and protracted negotiations between OPEC member governments and transnationals finally brought about nationalization in some countries and participation in others. This development and OPEC's far-reaching oil price decisions of 1973 and 1974 constitute the two most significant events in the history of the international petroleum industry. Transfer of complete ownership and control over petroleum resources in oil exporting developing countries from transnationals to state petroleum enterprises is now inevitable.

With the emergence of these profound structural changes in the petroleum industry, the traditional policies of transnationals regarding exploitation and development of petroleum resources of developing countries are bound to be replaced by policies which give equal weight to the aspirations of developing countries for national development and the long-term energy requirements of the world. A case in point is the current policy position of Saudi Arabia.

The prime objectives of Saudi Arabia in this regard include diversification of the sources of national income; development of an integrated national oil and gas enterprise, with special emphasis on development of national marketing capability to serve international oil and gas markets; increased processing of petroleum in Saudi Arabia and increased national participation in international energy transportation industry; augmentation of current oil and gas reserves and production capacities; and conservation of depletable oil and gas resources. Of equal importance is the development of alternative sources of energy and international cooperation, with special emphasis on assistance to and protection of developing countries' interests in the field of energy. These are the main factors that will determine the future role of Saudi Arabia's state petroleum enterprise in world petroleum and energy industries.

The events of 1973 and 1974 showed that the number of government to government oil transactions are apt to increase in times of supply shortages. A state petroleum enterprise is basically a government's business arm charged with implementing its national and international petroleum policies. With the expected future oil shortages, state petroleum enterprises are bound to play a highly significant role in such government to government oil sales. However, it would be incorrect to convey the impression that the rise in number of state petroleum enterprises at any stage could eliminate the need for the transnationals.

In fact their roles are complementary. Cooperation between state petroleum enterprises and transnationals could bring great benefits to all concerned, and there are a number of specific instances in which such cooperation is desirable.

Until the early 1970s, many experts felt that the supply of oil would continue to meet demand, with surpluses showing up from time to time. Also, for several decades the supply and availability position of the transnationals was strong enough to enable them to meet all demands whenever and wherever they occurred. This is not actually true anymore. It is now fairly well established that oil shortages may be witnessed in the 1980s, and that, in the future, transnationals may have insufficient supplies to meet their own integrated needs. The situation depends on many political and marketing factors. However, it is clearly evident that after the complete acquisition of petroleum producing facilities by governments, the supply and availability position of state petroleum enterprises in oil-exporting developing countries would be more stable, though not necessarily more abundant. It is also a fact that the transnationals, with their established integrated network and know-how, still command a strong position in the international petroleum market that provides a degree of stability and security in the flow of oil to its traditional markets. This factor may make it acceptable to the state petroleum enterprises in oil-exporting developing countries to leave sizable quantities of their crude available to be marketed through the transnationals. Under such conditions, the transnationals may continue to have more crude than the state petroleum enterprises, though certainly, without the same degree of security of supply.

Although many state petroleum enterprises have undertaken exploration programs of their own, the fact remains that exploration for hydrocarbons in developing countries is still largely a game for the transnationals. It, therefore, seems advisable for state petroleum enterprises to cooperate with transnationals in this field to the mutual benefit of both. That, of course, does not mean that the state petroleum enterprises may not play an important role in seismic, geophysical, and other surveys, as well as in drilling. In essence, it means risk taking and identification of exploration opportunities and drilling sites. It means interpretation of geological information, an activity in which the transnationals have proved their worth in the past and where they can continue to contribute to the benefit of all concerned.

With the completion of gas projects in a number of oil-exporting developing countries, supply of natural gas liquids and liquid propane gas is expected to surpass demand by a considerable margin for many years to come. Under these circumstances, transportation, marketing, and distribution require selective downstream investments, both by state petroleum enterprises and transnationals, in terminals, storage, transportation, and distribution facilities. Two major markets for these products are Japan and the United States. Similar facilities may be required in certain large consuming developing countries.

As a result of the work of the Energy Commission of the Conference On International Economic Cooperation (CIEC) held in Paris from 1975 to 1977, it has become abundantly clear that energy supply and demand balance in the 1980s and beyond cannot be achieved unless major energy projects are begun in almost all parts of the world, an undertaking which would require considerable international cooperation. State petroleum enterprises may play an important role by participating in the planning and implementation of an international energy development program. The oil-exporting developing countries will also need a different energy mix in their domestic markets in the not-too-distant future. The finite supply of oil and gas resources makes it imperative that these countries develop alternative sources of energy to meet their long-term domestic needs. For Saudi Arabia, nuclear and solar energies appear to be potentially suitable. Saudi Arabia has already signed agreements for joint solar energy research. Many of the transnationals have already extended their activities to other energy fields where state petroleum enterprises may prove to be excellent tools for the transfer of alternative technologies through cooperation with the transnationals and other international energy institutions.

The economies of the energy importing developing countries do not possess the ability to adjust easily to structural changes in the world economy, particularly those changes related to the new economic situation that add to the constraints on their development progress. Alleviation of these problems, therefore, requires special measures, international cooperation, and assistance. In the face of anticipated future oil shortages, it is imperative that state petroleum enterprises in these countries adopt aggressive programs to improve their future energy supply position. Such programs may include intensified exploration as well as expansion, development, and diversification of their indigenous hydrocarbon resources and related infrastructures. In addition, these countries may also participate in the exploration and development of hydrocarbon resources outside their countries with the cooperation of other petroleum enterprises. Ways and means must also be devised to ensure the efficient use of oil for essential transportation uses only, while rapid development and utilization of indigenous energy resources other than oil must be undertaken for nontransportation uses.

Because in most cases the capital and technology required for energy development are held primarily by the private sector, public financial institutions should play an increased role in financing energy exploration and development in developing countries, particularly those that import energy. New procedures should be introduced to enhance the complementary roles of all sectors in channeling flows of capital and technology into these countries.

The United Nations can play a vital role in this respect. In its final proposals, the CIEC made specific recommendations concerning the International Bank for Reconstruction and Development (IBRD), the International Development Association, the IBRD/International Monetary Fund Development Committee and other international and regional financial institutions. These recommendations need serious consideration and expeditious action.

The rise of state petroleum enterprises in developing countries to their present significant position in the international petroleum industry represents progress in the right direction toward the continuing quest of the developing countries for the creation of a New International Economic Order. Their future role is expected to bring about major advances in this direction.

5 State Petroleum Enterprises: Some Aspects of Their Rationale, Legal Structure, Management, and Jurisdiction

Hasan S. Zakariya

The world petroleum industry is now just over 100 years old. Its cradle was the United States, where the monolithic creature of today was born, reared, and grew into tempestuous manhood. Throughout the entire history of this crucial industry in the United States, private capital and initiative have been unchallengeably predominant. To reduce the question to its simplest elements, this was due mainly to the obvious fact that the free enterprise system has always been a cherished doctrine of political theory in that country and a preferred way of life. It was also due to the strange legal concept, now almost unique in the entire world, that the private owner of a piece of land is entitled to all the mineral riches that might lie underneath, all the way to the core of the earth!

Although the private petroleum companies in the United states have seldom been particularly popular over the years in their home country, they have managed somehow to ward off any attempt to curtail their essential jurisdiction or diminish their dominance. A few years ago, at the height of the most recent public controversy involving these companies, an attempt was made in the U.S. Senate to establish a state oil company to look after the interests of the community at large - public interests which do not necessarily always coincide with those of the private companies. As was widely predicted at the time, the move was doomed to failure. In fact, what was surprising about the move was not its failure, but rather its being made in that august body in the first place.

The story of the private oil companies in the United States which is, as is well known, long, dramatically rich and very fascinating indeed, has been told and retold over the years in countless pages. What has just been said is, of course, nothing but a tiny footnote, by way of introduction to the topic under discussion. With that particular purpose in mind, it is enough to point out that, as far as the petroleum industry is concerned, the basic role of public authorities in the United States has been merely that of a regulator of private companies and a licensor of petroleum rights to private interests on federal and state lands.

28

Outside the United States, the marketing of petroleum products (mainly of American origin), and the search for new sources of global supply began in earnest around the beginning of the present century. Again, private capital and private initiative were the predominant forces. The American private oil companies were the first to establish themselves abroad, not only to promote the sale of their own products, but also to look for new reserves to supplement, or even to replace, their domestic sources which, strangely enough, were thought at the time to be on the point of drying up.

Some of the industrialized countries of Western Europe also appeared early on the petroleum scene. Although they had little or no indigenous potential at the time, thanks to the colonial regime which was still in full force, these countries had vast areas of great potential under their control in various parts of the world. It is to be remembered that the governments of these countries brought the full impact of their political and military power to bear on the side of their private firms and entrepreneurs. The rest of the story is too well known to be retold here.

In later years, most West European countries, however, were inclined to intervene more directly and more authoritatively in the various activities relating to the petroleum industry. Apparently, these countries considered rightly that petroleum was, to borrow Clemenceau's famous dictum, too important a business to be left entirely to private interests. State intervention was rather modest and indirect at the outset; it represented the attempt to secure a foothold, for strategic and financial reasons, in such private entities as, for example, the Anglo-Persian (now BP) in 1914. In France, the government's intervention in oil increased over the years, resulting in 1924 in the creation of Companie Francaise des Petroles (CFP) to manage the German shares in the Turkish Petroleum Company (subsequently known as the Iraq Petroleum Company) by the San Remo Treaty of 1920. The French government acquired a 49 percent share in the CFP capital. Then the first purely state-owned company appeared on the scene in 1926, the Italian Azienda Generale Italiana Petroli (AGIP), whose task was to engage in petroleum exploration in Italy and abroad. There followed a number of other state-owned petroleum enterprises in various European countries, mainly after World War II.

The main reason for setting up these state petroleum enterprises was to act on the international scene as an instrument of implementing the national policy of their respective governments in order to ensure the security of foreign supply on the most favorable terms possible. As it happened, some of these European state enterprises proved to be the catalysts which, inadvertently rather than by original design, undermined the integrity of the old concession regime and helped the process of the emergence and the consolidation of state petroleum enterprises in the developing world.

In the last five years or so, a new breed of state petroleum enterprises made their appearance on the European scene, such as the Statoil of Norway and British National Oil Company (BNOC). As in the case of the state enterprises in the producing countries in the Third World, the main purpose of these European companies was to effect state participation in the indigenous and quickly expanding petroleum industry and not to look for possible secure sources of supply in foreign lands.

In seeking new sources of supply abroad - in Latin America, the Far East, the Middle East, and even in Czarist Russia and Rumania -the legal instrument for securing rights to the private American and European oil companies was the concession agreement. As the name readily denotes, the host countries had to concede to the foreign companies the exclusive right to search for petroleum, to produce it, own it, transport it, dispose of it in any manner they saw fit, and finally to fix its selling or transfer price. Those early concessions used to cover, more often than not, the entire territory of the country and were supposed to run for something like 60 to 99 years. Mandatory relinquishment of the concession area was still something unheard of in those days. All these far-reaching privileges were granted in return for a very modest reward; a royalty of a few shillings per ton of crude oil was practically the only payment.

Under the old concession regime there was virtually no role for the producing countries to play except to wait for the periodical payments to be made by the concessionaires. A minor office in some government department or ministry was more than adequate to cope with this limited passive role; to keep payment records and file routine reports. In those days, it was fashionable to refer to host governments, rather pejoratively, as "sleeping partners" of the foreign oil companies. Sleeping, certainly they were - partners, hardly! Things were destined to change, though, both drastically and dramatically.

The story of the state petroleum enterprises in the developing countries is essentially the story of the struggle against the old concession regime which, in turn, was a legacy of the colonial regime. For those who had the good fortune to become petroleum producers, the question was how to shake off the bondage of the early concessions and obtain a better deal. For the others, how to avoid falling into the trap from which the former had, to varying degrees, succeeded in extricating themselves.

In reviewing the story of the struggle against the concession regime and the emergence of the state petroleum enterprises, a special tribute must be paid to the Latin American countries which, as a group, deserve major credit for their pioneering efforts.

The most notable example is, of course, Mexico, where the nationalization of foreign concessions in 1938 was a major landmark on the long road to the reassertion of national control over natural resources. That nationalization was the culmination of three decades of controversy between the concession-holding companies and successive Mexican governments following the Mexican Revolution of 1910. Nationalization was, at the time, the only way out of the deadlocked situation where the companies adamantly refused to renegotiate their inequitable concessions and surrender some of their more excessive privileges. Petroleos Mexicanos (PEMEX) was the instrument for effecting that nationalization and receiving its enormous legacy. The story of PEMEX and its remarkable performance over the last forty years, in the face of great odds, is probably too well known to be retold here. However, that story has at least one basic lesson to teach:

The nationalization of Mexican oil provides Latin America's - and possibly the world's - outstanding example of action by a poor, undeveloped nation against what in 1938 was termed the international petroleum cartel. Politically, it showed that the power of this "cartel" could be contained by resolute action which, in the final analysis, the companies would have to acquiesce unless they could persuade the U.S. and Britain to intervene with physical force. Technically, it has demonstrated that a state oil enterprise, responsible for all aspects of oil operations from exploration through to final sales can, in the long run, operate an oil industry. And economically it has indicated how a nationally owned company can be organized and controlled by the State to meet what are considered to be required national objectives - which in Mexico's case have been increasing supplies of energy at low prices to the "consumer." (1)

Another example is Bolivia, which nationalized existing concessions a year earlier than Mexico (1937) and handed the properties over to Yacimientos Petroliferos Fiscales Bolivianos (YPFB), and kept all oil operations in the hands of this state company for 15 years. YPFB discovered new reserves, built refineries and pipelines, and achieved self-sufficiency in oil for Bolivia. Although the foreign companies were permitted, from 1952 onwards, to return to Bolivia to operate under terms different from those which were obtained under the old concessions, YPFB has maintained its monopoly over its own areas.

Argentina can, of course, boast of creating the first state petroleum company in Latin America. Yacimientos Petroliferos Fiscales Argentinos (YPF) was founded in 1922, and although it was not given at first any formal monopoly of oil in the country, it was destined to play a much more decisive role in later years. From the mid-1930s, the government began to restrict the activities of the foreign private companies, and by mid-1958 the oil policy was altered. YPF monopoly of oil development in Argentina was formally confirmed, but YPF was at the same time authorized to make arrangements with private oil companies to explore, develop, and produce oil under contract to it, a step which proved to be the forerunner of a new type of petroleum agreement in the developing countries known as the service or production-sharing contract. Some were simply drilling contracts at a fee per meter drilled, under which YPF was entitled to the whole value of the oil. Others were classed as development contracts, under which the private companies were required to sell any oil eventually produced to YPF at prices related to world market levels, thus giving the contractor a return on his investments. A third type, which involved exploration as well as development, provided for recovery of the foreign risk capital through a percentage of the value of any crude produced, plus a reward of a given percentage on all oil produced over a period of time. Under this form of contract, title to the oil was retained by YPF. (2)

In 1938, Brazil declared the petroleum industry a public utility and placed it under the direction of a National Petroleum Council. After the 1950 presidential election, which was fought under the slogan O petroleo e nosso (petroleum is ours), more comprehensive oil legislation was enacted forming Petroleo Brasileiro (Petrobras) a state-owned company with monopoly rights for exploration, production, and all new refining development.

Venezuela, although until recently epitomizing the gradualistic approach of successive revisions and readjustments instead of drastic and swift change, had persistently endeavored, with no small measure of success, to regulate and control the activities of its concession-holding companies and to maximize oil benefits.   Thus, as early as 1943, Venezuela was able to renegotiate its concession agreements and to obtain more favorable terms which other producing countries were not able to do until the 1950s and 1960s. There are, of course, other state petroleum enterprises in Latin America, both old and new, whose performance is worthy of note. Because of the limited scope of this introductory review, one can only mention them in passing: ANCAP of Uruguay, CEPE of Ecuador, ECOPETROL of Colombia, ENAP of Chile.

In the Middle East and North Africa, the emergence of state petroleum enterprises was, as in Latin America, a direct logical result of the move to eliminate or, at least restructure, the concession regime.   Bearing this in mind, it was only natural that most state petroleum enterprises in this group of countries were established after OPEC came into being in 1960. One major state enterprise in this area owes its existence, however, to a major event which predates the creation of OPEC by almost ten years. This is NIOC of Iran, set up in 1951 upon the nationalization of the Anglo-Persian concession. Although well-known subsequent developments tended to frustrate the effects of that historic measure and to cancel out some of its objectives, NIOC was, nevertheless, able to live on and to grow steadily in stature.

All OPEC countries, except one, now have their own state petroleum enterprises; the only exception, Gabon, is actively considering setting up its own. The main reason for the creation of the state petroleum enterprises within the OPEC community of nations was to help achieve state participation in new petroleum agreements with newcomers in areas either not previously committed to existing concessions or relinquished, voluntarily or by force of law, from these concessions. This took the form of joint-ventures and service contracts. We shall have occasion to discuss this matter in more detail later.

Another important consideration for establishing state petroleum enterprises within the OPEC community was in anticipation of introducing state participation in the existing major concessions - an issue which was quite alive in certain member countries even before OPEC formally endorsed it in its Declaratory Statement of Policy of 1968 (Resolution no. XVI-90). State participation in existing concessions finally came about in 1971, first at a negotiated minority rate of 25 percent, and soon afterwards at a majority of 60 percent. Those OPEC countries which did not want to wait until the early 1980s to eliminate completely their concessions, opted instead for a full state take-over, as did Algeria and Iraq in 1971 and 1972 respectively (by a sovereign act), and Venezuela in 1975 (by a negotiated deal). Some of the others who initially chose state participation, like Kuwait and Qatar, shortened the process a few years later and also effected a complete take-over. In most of the above instances, the state petroleum enterprises were established and ready to serve as an effective instrument for implementing state participation in, or complete nationalization of, the existing concessions and to act as the natural depository of the new rights and duties arising therefrom.

Outside Latin America and the OPEC community, where the existence of a state petroleum enterprise has become the general rule, many developing countries can now also boast of their own state petroleum enterprises. Some of these are as well-established and sophisticated as, for example, the Indian Oil and Natural Gas Commission (ONGC) which was created in 1956 and has been successful on its own, both within India and abroad. The Egyptian General Petroleum Corporation (EGPC) is another notable example of a state petroleum enterprise which has a long, successful record of exploration and development of petroleum, often in partnership with international companies under nonconcessionary types of agreements such as the joint-venture and production-sharing. A more recent national entity is the Malaysian State Petroleum Company, Petroliam Nasional Berhand (Petronas), which was set up under the Petroleum Development Act of 1974. Petronas, vested with the entire ownership of all petroleum resources in Malaysia, was instrumental recently in converting all existing concessions in that country into production sharing contracts and thus achieving its first and major objective.

Most state petroleum enterprises in the developing countries have assumed the legal form of public corporations. As such, their structure, functions, management, and accountability have been decisively influenced by the general theory and practice of the "public corporation" as developed during the last hundred years or so in the legal systems of the industrialized countries of Europe and North America. Consequently, it is deemed useful, if not essential, to begin this part of the discussion with a brief review of the development of that general theory. It is believed that, in addition to its intrinsic informative value, such a review would help the critical appraisal of the constituent statutes of some of these entities by demonstrating to what extent these statutes have conformed with or departed from the well-established trend elsewhere.

Traditionally, the state has fulfilled its basic functions of collecting taxes, ensuring public order and national security, administering justice, and other related matters through departmental administration. In modern times, however, as the state was called upon to fulfill ever increasing functions of a predominantly social and economic nature, it appeared that the conventional government departments were not adequately equipped to handle such new functions which, generally were not of a very specialized and highly technical nature. Besides, the inherent defects of the departmental administration - the bureaucracy -

had been under constant attack for some time.   This distrust of "government by civil service," was also a significant factor in the emergence of a policy of public administration through separate agencies that would operate largely according to business principles and be separately accountable. (3) Whether the distrust of bureaucracy is fully justified or exaggerated is quite another matter, and is, naturally, beyond the scope of this paper. However, it would seem, on balance, that the autonomy, or lack of it, of governmental agencies is a relative matter since, as is rightly pointed out, "No public corporation is completely autonomous. No government department administering a service of some magnitude can be managed without a considerable degree of independence. It is a matter of degree, or shades which vary according to political considerations, to the traditions of the country concerned, the efficiency of the civil service and a good many other factors." (4)

These two factors resulted, among other things, in the new state functions - overwhelmingly social and economic in nature, as we have said - being diverted and channeled through new entities. They are mainly of two types: joint stock companies controlled completely or partly by the state; and public corporations.

Some countries (such as Austria, Finland, Greece, and Sweden) use the joint stock company as the main form of government economic activity. In others it is regarded as the most convenient instrument for government participation in joint ventures with private interests. Yet, in certain others, like the United Kingdom, the commercial company is not considered the most appropriate form of public enterprise, and they have opted instead for the public corporation. France is believed to be the prime example where the commercial company is resorted to as frequently as the public corporation proper. In France, the former has been retained first for existing commercial companies in which the state has become the sole shareholder (as in the case of the Banque de France and the nationalized insurance companies), and, second, where the state has acquired a controlling interest in commercial undertakings without displacing private share holders (such as the railways and Air France). (5)

As far as state petroleum enterprises in the developing countries are concerned, the commercial company has not been widely used. NIOC of Iran seems to be the most notable example of this particular form; its statute of 1960 declares at the very outset that NIOC is "a commercial joint-stock company and shall carry out its operations in accordance with the provisions of these statutes and the Commercial Law." However, NIOC shares belong entirely to the government and are nontransferable.

Another example is the old Kuwait National Petroleum Company (KNPC) which, as a share company, was established in conformity with the Commercial Company Law; its capital was subscribed to at the rate of 60 percent by the government and 40 percent by private interests. A few years ago (by Law no. 8 of May 19, 1975), however, the government bought out the entire private share and thus became the sole shareholder of KNPC.   A new statute for KNPC was issued on November 6, 1977.

The public corporation is today a worldwide phenomenon. There are very few states in which the public corporation does not play a significant, indeed an indispensible, role in its economic life. It exists in such widely divergent political systems as those, for example, of the Soviet Union and the United States. In the USSR, state trusts, which are semiautonomous government corporations with a separate legal personality, were established as early as 1923. In the United States, public corporations or authorities are in charge of certain important economic activities. On the federal level, the most outstanding example is the Tennessee Valley Authority (TVA), set up in 1933. On the state level, there are numerous public corporations which are active in the field of public utilities, transportation, and communication.

In the United Kingdom, the appointment of ad hoc bodies has often been the preferred method of dealing with particular problems which have arisen. While examples of such bodies can be found from early times, the present practice of the public corporation as a form of government administration is said to date from the first half of the nineteenth century. (6) More recently, in the middle of the present century, a great surge in the number of public corporations took place in the United Kingdom as a result of the major nationalization program of coal, electricity, gas, and transport.

The emergence and growth of the public corporation reflect a radical change in political theory as to the proper functions of the state. Until the nineteenth century, the classical theory was that the state had no business in the field of commerce, industry, or social welfare. As is well known, this rather primitive attitude could not be maintained very long; it is universally recognized now - albeit to varying degrees - that the intervention of the state in these fields is not only legitimate and appropriate, but even essential at times. Under the dual pressure of these new and increasing functions and the growing technical complexity of many of them, the inadequacy of the conventional type of public administration became, as already indicated, more and more evident, and a new formula had to be found. This new formula was the public corporation, somewhere midway between a purely government department and a private commercial company. In his message to Congress in 1933 recommending the formation of the Tennessee Valley Authority, President Roosevelt eloquently referred to the new formula as a "....corporation clothed with the power of government but possessed of the flexibility and initiative of a private enterprise."

In spite of the diversity of background, economic and political objectives, and prevailing legal systems, public corporations established in different countries possess, by and large, certain common features. By way of illustration, public corporations as pioneered and developed in the United Kingdom and the British Commonwealth are characterized by the following: (1) they are separately established by statute; (2) each has a separate legal personality; (3) their administration is in the hands of a governing board appointed by the government; (4) their employees are not civil servants; (5) the basis of their finance is not parliamentary appropriation but permanent revenue earning assets (some of them, with ministerial approval, can raise loans; they have to balance revenue and expenditure and any profits generally have to be ploughed back into the development of the enterprise; they are commercially audited); (6) they are responsible to the government through the appropriate minister and subject to his general direction; (7) in their day-to-day operations they are like other private legal entities; they are fully liable in law and do not enjoy any of the legal privileges and immunities of government. On the whole, public corporations are said to have a dual nature. In their commercial and managerial aspects they resemble commercial companies and they have an essentially private law status. But insofar as they fulfill public tasks on behalf of the government, they are public authorities and, as such, subject to control by the government, within the limits defined by statute. (7)

Public corporations as developed in other countries also have general features similar or corresponding to the above. In France, for example, where they are known by the term establissement public, their dual nature is again distinctly clear. While they resemble private entities in their outward legal form, and conduct their activities like commercial enterprises, the public nature of their services demands that the state subject them to its control and that they are, in part at least, governed by public law. (8)

Since their creation was certainly influenced, if not modeled, on the well established examples in the industrialized developed countries, it was only natural that state petroleum enterprises in the developing countries should assume, with some notable variations, most of the above-mentioned characteristics.

The administration of public corporations is normally entrusted to a semiindependent governing board appointed by the executive power. Even though it commands a certain measure of autonomy in managing the affairs of the corporation, the board is, at the same time, subject to public control and accountability. The governing boards are usually composed of qualified members of suitable experience and expertise. In the United Kingdom, for example, the chairman and members of the boards are appointed by the appropriate minister who, even though he has a very wide discretion in this matter, is required in some instances "to appoint persons appearing to him to be qualified as having had experience of and having shown capacity in industrial, commercial or financial matters, applied science and the administration or organization of workers." (9) In France, on the other hand, appointment to the board of the new public corporations is based on a tripartite system of representation: government, trade unions, and consumers' organizations. (10)

The statutes of the state petroleum enterprises in the developing countries generally adhere to the principle of autonomy. They are also governed by a semiindependent board of directors. In the words of the Statute of Petromin (Saudi Arabia), for example, the board "is the authority that controls its (Petromin's) affairs and the conduct of its business and lays down the general policy to be followed, without being bound by the administrative and financial regulations applicable to Government departments." A more elaborate statement of this principle of autonomy is found, for example, in Article XI of the Law no. 123 of 1967 establishing the Iraq National Oil Company (INOC):

> The management of the Company and the attainment of its objectives shall be undertaken by a financially and administratively independent Board of Directors. The Board shall exercise all the powers and rights vested in the Company under the laws in force,.... The Board shall draw up the Company's policy in all fields including administration, accounts, production, marketing on a sale or barter basis, and project implementation. It shall supervise the execution of such policy.... The Board shall decide upon the Company's organizational structure, divisions, departments and sections at its headquarters and elsewhere and may delegate powers and authority, as it deems fit, to the President, Managing Director and General Manager.

Members of the board are appointed by the government (normally the Council of Ministers), the sold shareholder of the state company. Some of them, understandably, are high officials serving ex-officio. Normally, they are identified in the statute right from the outset, and belong to ministries or departments to which the activities of the state company are related. Others are appointed from outside government service, upon nomination of the appropriate minister. Some statutes specify the kind of background and expertise of such outsiders; others leave the matter entirely to the discretion of the minister.

Chairmanship of the board varies from one statute to another. Some assign the chairmanship to the appropriate minister or his representative (Abu Dhabi, Saudi Arabia, Venezuela). Others assign it to someone other than the minister or his representative (Iran, Libya, Indonesia).

The public corporation is, of course, nothing but an arm of the state, which creates it and owns its entire capital. Even though it is widely recognized that the public corporation should enjoy a measure of independence (financial and administrative) for the sake of better fulfillment of its vital objectives, it is equally recognized that its activities should be subject to the scrutiny and control of higher authorities. After all, he who pays the piper calls the tune! Such control is deemed essential, not only to ensure that the public corporation is faithfully adhering to the policy directives of the government, but also that it is not transgressing the limits of its jurisdiction as laid down in its statute and other applicable laws. It is admitted, however, that this dialectical interaction between the two opposing needs - autonomy and control - can sometimes be very difficult to regulate in practice; the right balance between the two can, and does indeed, vary from one case to another.

State control of the public corporation is usually exercised through the appropriate minister to whose ministry the corporation is hierarchically attached. The degree of ministerial supervision varies, of course, not only from country to country but also from one public corporation to another. In the United Kingdom, for example, the appropriate minister is entitled to give certain public corporations directives of a general nature as to the exercise of these functions in relation to matters affecting national interest. He is required, though, to consult with them before giving any directives. (11) There is also a general obligation on the part of most British corporations "to furnish the Minister with information which he requires, to submit returns and accounts to him, and to provide him with facilities for verification of the information in such manner at any such times as he may require." (12) Most corporations are required, after the end of the financial year, to report to the appropriate minister on the exercise of their functions. The minister must lay the report before Parliament. (13)

While the above indicate unmistakably that in the United Kingdom the minister's power in this regard is considerable, in practice, his actual power is said to be greater than his statutory power. "His influence and his power to give directions are more important than the directions themselves, indeed the fact that he has the power is often sufficient by itself." (14)

As the preceding paragraph would suggest, government control of public corporations can assume several different forms: it can be a priori or a posteriori; it can be exercised from within or from without; it can be open and direct or subtle and discreet.

The a priori control can sometimes be exercised, for example, through the appointment of the members of the board of directors. If all or the majority of the board are drawn from various government departments to serve ex officio, the likelihood of their susceptibility to pressures from above or their eagerness to avoid any possible dispute with other departments is perhaps stronger than if the majority is drawn from outside government circles. This rather indirect, and subtle, type of control (which is not infrequently exercised) may also be termed as control from within.

In this connection, another related and rather more crucial question should be considered: How best to ensure ministerial supervision and control? Is it necessary or desirable that the minister or his representative be a member of the management board? Even though the practice varies, it would seem that the general trend is against this proposition. On this point, countries which are as different as those of the United Kingdom, the Federal Republic of Germany, France, the USSR, the United States, India, Italy, and Australia have all been in favor of excluding the appropriate minister from the management boards of their respective public corporations. Canada appears to have been heading in the opposite direction. (15) It is, of course, recognized that in mixed ventures - the joint stock company - the representation of the state by civil servants, or, in some cases ministers, is inevitable. But in the case of public corporations, the blurring of managerial and supervisory functions is not deemed desirable; indeed, it may well infringe upon the basic principle of the autonomy of public corporations.

Let us turn now to the specific case of the state petroleum enterprises in the developing countries. In reviewing their various statutes, it can readily be concluded that they, too, are subject to the same rule of state control and generally in the same manner, at times even to a larger extent and with greater force.

State control of state petroleum enterprises is also exercised by an appropriate minister to whose ministry the company is hierarchically attached. This is usually the ministry of petroleum, and if there is no such separate ministry, then to the ministry which has jurisdiction over petroleum matters, such as the ministry of mines, of industry, of finance, or of national economy.

As far as state petroleum enterprises are concerned, the practice is not uniform as to whether or not the appropriate minister or his representative is included in the management board. In either case, the supervisory power of the minister does not seem to amount, in theory at least, to an outright veto power. Generally, decisions of the board of directors are considered effective upon issuance, unless the minister objects to them and returns them to the board for reconsideration, or except where the statute provides for mandatory prior approval by the government. Whether or not the minister (or his representative) is a member of the board, the statutes of some state petroleum enterprises have provided for some orderly procedure for settling disagreements which might arise between the minister and the board.

In a case where the minister is the chairman of the board, such as in Petromin of Saudi Arabia, a two-third majority of the board can overrule him (Art. 7 of the Statute). This is a built-in mechanism for overcoming a potential problem. In a case where the chairmanship belongs to someone other than the minister or his representative, as in NIOC (Iran), LNOC (Libya) etc., the council of ministers is made the final arbiter between the two parties. This procedure also applies in Abu Dhabi, even though the minister there is the chairman of the board. In Iraq, the President of the Republic, or whom he may authorize, is the final arbiter in such cases (Art. 4 of Law no. 76 of 1971).

The above notwithstanding, it is very doubtful whether such a statutory mechanism for solving disagreements between the appropriate minister and the state petroleum enterprise has frequently been resorted to in practice. It would be very interesting indeed to learn whether, how often, and in what kind of dispute this procedure has actually been put to the test.

Some statutes require the submission of certain decisions to the council of ministers for approval before they can become effective. Thus, Article 8 of the Statute of the Venezuelan Petroleum Corporation (CVP) of 1966 provides, for example, that the Board of CVP shall submit, among other things, the annual budget of the company to the National Executive (council of ministers) for approval or modification. The Law no. 7 for 1971 establishing the Abu Dhabi National Oil Company provides in Art. 14 that the decisions of the company's board of administration concerning involvement of the company in exploration and drilling for petroleum, or establishment of companies on its own, or in association with others or participation in existing companies, must be submitted to the council of ministers for approval before they can be valid. Similar provisions can be found, for example, in the statutes of INOC (Art. 15) and LNOC (Art. 13(4) and Art. 19).

One of the main objectives of the creation of a state petroleum enterprise is, of course, to promote the speedy exploration and development of petroleum deposits under effective national control and to maximize the financial benefits from their eventual exploitation. To that end, state petroleum enterprises are usually granted, either in their statutes or in petroleum laws, primary and exclusive jurisdiction over all the national territory, including the continental shelf, with the exception of those areas, if any, which have already been committed to foreign operators under existing agreements. In addition to that, if state participation in the equity of such existing agreements has been achieved or is contemplated, such participation is usually also assigned, with all its underlying duties and privileges, to the state petroleum enterprise. The same also applies when existing foreign concessions are nationalized through a sovereign act of the state or completely taken over by virtue of a negotiated deal. The state petroleum enterprise is the natural instrument for effecting any such state take-over and of receiving its legacy. Sometimes the legacy can be so varied and complicated and of such magnitude that the state finds it advisable to create one or more special state petroleum enterprises, under different appropriate names, for that purpose. This is what happened, for example, in Iraq in 1972 upon the nationalization of the IPC concession when a new state company was established; and, more recently, in Venezuela where several state-owned corporations were set up, in addition to the existing CVP, to implement the nationalization of the several major concessions at the end of 1975 and to continue with normal operations.

While it is the general, and indeed the preferable, practice to grant the state petroleum enterprise exclusive territorial jurisdiction right from the date of its inception, this is, by no means, the rule everywhere. Some statutes have deferred this matter to the future to be decided upon by the government which may well decide to allocate to the state petroleum enterprise only a few areas on a gradual and selective basis. A notable example of this course of action is to be found in the Law no. 3 of 1968 establishing the state petroleum enterprise in Libya, which provides: "The Corporation shall have the right to exploit areas allocated to it either on its own or in participation with others. The allocation and demarcation of such areas shall be fixed by a decision of the Council of Ministers" (Article 3-1).

While there is no doubt that endowing the state petroleum enterprise with primary and exclusive jurisdiction is a logical step, in regard to which there is (and should be) no serious reservations or second thoughts, it is to be borne in mind that, in exercising that jurisdiction, the state petroleum enterprise does not usually enjoy full and unfettered discretion. Otherwise, this would contravene the basic premise that the state petroleum enterprise is merely an agent of the state and, as such, must faithfully conform to the broad policy guidelines laid down in its statute or other relevant legislation. Let us cite a few examples.

When a state petroleum enterprise is established to effect a policy of state monopoly of exploration and production, it is naturally enjoined from seeking outside operators and, therefore, must carry out its task alone with its own technical and financial resources. The most obvious example for this absolute injunction is, of course, those Latin American countries such as Mexico, and until the rule was relaxed recently, Brazil, Argentina, and others.

Sometimes the injunction is only a relative one, and pertains only to the types of petroleum contract or agreement which the state petroleum enterprise is not allowed to enter into or, conversely, the types which it may enter into. The following examples would serve to illustrate this:

1. The National Iranian Oil Company (NIOC) had been granted primary and exclusive jurisdiction (with the exception of the area of the foreign consortium) by virtue of Art. 5 of the Iranian Petroleum Act of 1957 and Art. 4 of its statute of 1960. NIOC was authorized to divide the territory of the country into petroleum districts and to declare any such district or part of it open for petroleum activities. While the right of NIOC to conduct such activities on its own is readily recognized under both legislations, it was not anticipated that NIOC would be in a position soon to shoulder all the financial risks and technical requirements alone. Consequently, NIOC was wisely authorized to seek the help of outsiders and to conclude petroleum agreements with them. Although there was no explicit injunction against any particular type of agreement in either legislation, it was hardly conceivable that NIOC (born out of the bitter controversy over the nationalization of the Anglo-Persian concession in 1951) would freely opt for the concession system in dealing with outsiders. Besides, the whole tone of both legal instruments is oriented towards the joint venture and similar nonconcessionary types of agreement. Under the Iranian Petroleum Act of August 6, 1974, however, the situation became unequivocally clear. This law provides that: "...The National Iranian Oil Company may negotiate with any person, whether Iranian or foreigner, and may prepare and execute any contract which it may deem appropriate on the basis of Service Contract and in conformity with the provisions and stipulations of this Act." (para. 2)

2. In Iraq, the injunction against the concession type of agreement was outright and fully pronounced. The Iraq National Oil Company (INOC) was assigned, by law no. 97 of 1967, the exclusive right to develop petroleum resources in all the Iraqi territory including territorial waters and the continental shelf. While INOC was required to develop the assigned areas directly,

or in association with others if it deemed this more conducive to the fulfillment of its objectives, it was categorically forbidden from entering into "a concession or the like" in dealing with foreign oil companies. (Art. 3-3)

3. There are several similar examples. One may recall in this connection the policy declared in Venezuela in the 1960s to the effect that there should be no more concessions and that only service contracts would be contemplated by the state company, CVP. One can also recall the Algerian Petroleum Law of 1971 which provides that no foreign company can engage in the search for and development of petroleum in Algeria without participation of the national company Sonatrach, which should own at least 51 percent of the joint venture.

Although in some instances there was no such specific injunction against, or preference for, any particular type of agreement, still certain other, and perhaps less stringent, conditions were laid down. This case is exemplified by the Law of 1968 establishing the Libyan National Oil Company (LNOC). Article 3 of the Law allowed LNOC to "exploit areas allocated to it either on its own or in participation with others." However, the Law provides that participation contracts to be concluded by LNOC must contain terms and conditions which are more favorable to LNOC than those already provided for under the Libyan petroleum law.

In this context it is to be noted that due to the paramount importance attached to the partnership contracts to be concluded with foreign firms and the understandable concern that such agreements should meet the prescribed standard, both in form and substance, it is required in all laws and statutes cited above that no such contract can be deemed valid without the approval of the council of ministers. This is still another legitimate manifestation of state supervision and control of the activities of state petroleum enterprises.

In order to fully achieve the basic policy objectives for which it was created, the state petroleum enterprise must be allowed the widest range possible of petroleum-related activities to be carried out either directly or through one or more subsidiary. In this respect, a greater measure of centralization would generally seem to be more conducive to efficiency and conservation of talents and financial resources than a plurality of parallel and unrelated entities.

In fact, most state petroleum enterprises have been granted a wide range of functions and are not confined to one or a few specified aspects of the petroleum industry. This wide range would usually encompass not only all the normal stages of the development of indigenous petroleum resources, but also all modes of transportation, storage, refining, trade, distribution, and petrochemical industires. To that end, state petroleum enterprises are normally authorized to set up branches, agencies, and subsidiaries, both within and outside the country, alone or jointly with other parties (Libya, Art. 9; Iran, Art. 2; Abu Dhabi, Art. 4; Iraq, Art. 3 of Law no. 97 of 1969; etc.).

In consequence, some state petroleum enterprises which had a modest beginning have grown in recent years into massive and fully integrated organisms. When it was set up in 1965, Sonatrach of Algeria, for example, employed less than 500 persons. With its subsidiaries, it employs now 80,000 persons. This was due to the fact that Sonatrach has never ceased growing over the years and its activities have become increasingly extensive and increasingly complex. (16)

The state petroleum enterprise, of course, can and does act merely as an operating company, as both operating and holding company, or just as a holding company. The Nigerian National Oil Company (NNOC), for example, is said to be a holding company, responsible for determining the policies of its several subsidiaries which are actually in charge of the various aspects of the operations. Each subsidiary has a representative on the board of directors of NNOC. (17)

As far as the activities of the state petroleum enterprise outside its own territorial jurisdiction are concerned, there are a few interesting cases, with varying degrees of success, in which some relatively old and well-established state petroleum enterprises have actually been engaged in recent years in petroleum exploration outside their home country. To give some examples: The Oil and Natural Gas Commission of India (ONGC) in Iran, Syria, and Tanzania; NIOC of Iran in the North Sea; YPF of Argentina in neighboring Uruguay; Petrobras of Brazil in Colombia (where it is already producing), Algeria, and Iraq (where it has made significant discoveries), Egypt, Libya, and the Philippines. There are similar ventures in the refining operations.

While it is not within the strict scope of this chapter to review in full this extraterritorial aspect of the activities of some state petroleum enterprises and to analyze their pros and cons, it may be mentioned in passing, that, at least as far as cooperation between the developing countries is concerned, such extraterritorial ventures should generally be viewed as positive and commendable.

Cooperation between state petroleum enterprises in developing countries and foreign firms, whether state owned, privately owned, or mixed, and the modalities such cooperation assumed in the last 20 years or so, has been one of the most crucial and dynamic aspects of the activities of state petroleum enterprises, particularly in the field of exploration and development of indigenous petroleum deposits.

It is to be recalled that in conferring primary territorial jurisdiction on their state petroleum enterprises many developing countries have, in principle, required them to carry out the petroleum operations directly on their own, if possible. At the same time, they also authorized them to seek the assistance of outside operators if this would be conducive to more effective performance. As it happened, this outside assistance has, on the whole, been so far the rule rather than the exception.

It is to be recalled also that, in regulating their relationship with outside operators, state petroleum enterprises have been enjoined, by and large, from adopting the concession type of agreement. This injunction is, of course, essentially sound, not only because of better past experience with the concession regime but, more importantly, because of the fact that this regime is basically a negation of the very raison d'etre of the state petroleum enterprises and an abdication of their role. The concession regime having been categorically excluded, the scope and modes of equitable and fruitful cooperation with foreign forms are nevertheless vast and varied.

Even when the concession regime was still essentially intact in the 1950s and 1960s in most of the producing countries in the developing world, state petroleum enterprises of some of these countries were beginning to experiment with new arrangements of a nonconcessionary nature. Part of the credit in that respect should, of course, be given to certain European state-owned companies and even to some of the privately owned ones, such as the American independents and the Japanese. These so-called newcomers, of course, had their own special motivations and were not acting for purely altruistic reasons. Nevertheless, the fact that they were willing to propose or accept new legal arrangements that were regarded at the time with great dismay as anathema, if not sacrilegious, by the major concession-holding companies, was very commendable on the part of these newcomers.

If any special mention is to be made in this connection, then this should go to the Italian state enterprise (ENI) which probably was, more than any other international oil company, responsible for making this new trend possible. Thus we find, as early as the mid-1950s, state petroleum enterprises like EGPC of Egypt and NIOC of Iran concluding joint-venture agreements with ENI, under which the state petroleum enterprise of the host country was assigned a partner role which could seldom be found in the major concessions of the day. Subsequently, several other state petroleum enterprises, mainly within the OPEC community, such as Petromin, KNPC, and LNOC, entered into a similar type of arrangement, not only with ENI, but also with other European state companies such as Aquitaine, Hispanoil, and others.

Another form of cooperation between state petroleum enterprises and foreign operators was that of the service contract and its main variant, production-sharing. This type of arrangement was first introduced in the Middle East by the two more or less identical agreements between the French state enterprise ERAP and NIOC and INOC, in 1966 and 1967 respectively. Elsewhere, it was also used by CVP of Venezuela and has been the standard form for all offshore contracts in Indonesia. In the 1970s the service contract, and particularly its main variant, production sharing, has become increasingly popular in developing countries due not only to its intrinsic simplicity and its balanced allocation of benefits, but also to its harmony with the basic precepts of state sovereignty over natural resources. EGPC of Egypt, for example, has concluded about 40 different agreements of this kind during the four-year period, 1973 to 1977. It may be recalled that the Petroleum Act of 1974 of Iran specified that the service contract is the only type of agreement into which NIOC may enter with outsiders. Other examples include among others Bangladesh, India, and the Philippines. (18)

Their large concessions have been recently abolished altogether, or radically altered through the introduction of state participation and

other measures. The major oil companies are now eager, as their minor
colleagues on the international scene, to operate as mere partners or
contractors to the state petroleum enterprises, both in developing
countries and elsewhere.

It is to be recalled that in some developing countries, particularly in
Latin America, the reaction against the old concession regime was so
bitter and so severe that some of these countries were anxious not only
to abolish that regime altogether and rid themselves of those operating
under its mantle, but also to establish an absolute state monopoly over
all matters relating to petroleum. In consequence, the state petroleum
enterprise in those countries were forbidden to entrust any of their
basic tasks to foreign companies.

The extent to which these countries were able to maintain the
doctrine of state monopoly varies of course. Mexico is probably the
prime, if not the only, country which has somehow managed to adhere
to that doctrine with remarkable steadfastness until this very day. The
major new reserves discovered by PEMEX a few years ago may have
been a contributing and welcome factor in this regard, since this came
at that particularly critical time when some less fortunate countries
were beginning to have second thoughts about the continued wisdom of
state monopoly in the face of dramatic increase in the price of foreign
oil and the questionable future availability of foreign supply.

Faced with crushing bills for imported oil and staggering costs of
operating in deep water areas, most of the others seem to have come to
the conclusion that the new and unforeseen circumstances called for a
more pragmatic approach. Consequently, state monopoly was somewhat
relaxed and foreign oil companies were invited to participate with the
state petroleum enterprises in searching for new indigenous reserves
under a well-defined set of terms. In this regard, one can cite several
examples such as Argentina, Brazil, Burma, and India.

Whether or not to allow the state petroleum enterprise to engage
the services of foreign operators as partners or contractors is, of
course, a matter of national policy to be determined by each developing
country in the light of its own circumstances at any given time.
However, it is evident that engaging the services of a foreign oil
company is not wrong per se, or derogatory of the principle of state
sovereignty over natural resources. It is the mode and conditions of
such cooperation that call for a careful scrutiny and painstaking
appraisal. If the conditions are right, such cooperation can be positive
and profitable to all concerned. Most, if not all, the developing
countries appear to have no qualms now as to the desirability, indeed
the necessity, of such positive cooperation. (19)

NOTES

(1) P. Odell, "The Oil Industry in Latin America", in E. Penrose, The Large International Firm in Developing Countries (Westport, CT: Greenwood, 1968) p. 289. See also Odell, "The Oil Industry in Latin America: Relationships Between the International Oil Industry and National Interests in Latin American Countries", paper presented to the OPEC Seminar on "International Oil and the Energy Policies of the Producing and Consuming Countries," 1969.

(2) J.E. Hartshorn, Oil Companies and Governments, (London: Faber and Faber, 1967), pp. 250-51.

(3) W. Friedman, ed., The Public Corporation - A Comparative Symposium, (Toronto: Carswell, 1954), p. 548.

(4) Ibid., p. 551.

(5) R. Drago, "The Public Corporation in France," in Hartshorn, Oil Companies and Governments, p. 127.

(6) J. Griffith and H. Street, Principles of Administrative Law (London, Pitman, 1959), p. 271.

(7) W. Friedman, "A Theory of Public Industrial Enterprise", in A.H. Hanson, ed., Public Enterprise - A Study of its Organization and Management in Various Countries (Brussels: International Institute of Administrative Sciences, 1955), p. 21.

(8) Drago, "The Public Corporation in France," p. 119.

(9) Griffith and Street, Principles of Administrative Law, p. 276.

(10) Friedman, "A Theory of Public Industrial Enterprise," p. 21.

(11) Griffith and Street, Principles of Administrative Law, p. 279.

(12) Ibid. p. 280.

(13) Ibid.

(14) Ibid. p. 293.

(15) Friedman, The Public Corporation, p. 562.

(16) K. Adeniji, "The Legal Anatomy of OPEC State Oil Corporations: Perspectives on Nigerian Oil Corporation," 8 Eastern African Law Review, no. 1 (1975), pp.

(17) Arab Oil & Gas (Paris), quoting Mr. A. Ghozali, President of Sonatrach, vii, no. 152 (January 16, 1978), 4.

(18) For a detailed review of the characteristic features of both the joint venture and the service contract see H. Zakariya, "New Directions in the Search for and Development of Petroleum Resources in the Developing Countries, 9 Vanderbilt Journal of Transnational Law, no. 3 (summer 1976): 545-77.

(19) In a recent interview, Mr. A. Ghozali, Algeria's Minister of Energy
   and Petrochemical Industries, and President and General Manager of
   Sonatrach, reendorsed this trend by saying "Sonatrach is anxious to
   undertake joint venture exploration with foreign companies, on the
   one hand because they contribute financing and, on the other,
   because it is useful for Sonatrach to take advantage of these
   companies' technical know-how.   We shall continue to develop
   associations with our present partners and with others." Arab Oil
   and Gas, p. 5.

6 Activities and Technical Capabilities of State Petroleum Enterprises

Sammi Sherif

The long history of the oil industry, stretching over a century, can be divided into three stages. Before the middle of this century the oil sector in the developing countries was paid no attention by host country governments, and the major oil companies had a free hand in most activities. During the period 1950 to 1970 some oil activities developed as part of the government determination to reach more equitable and contractual terms with the powerful transnational petroleum companies. As part of state resolutions to assert their rights, some state petroleum enterprises were created, but they remained inactive or concerned themselves with concessionary matters only. Later on, there was an escalation of the differences between the transnational oil companies and the state petroleum enterprises, independently or united together under such organizations as OPEC, OAPEC, ARPEL, and others. This was a result of the constant effort of national authorities to gradually achieve control over oil and other natural resources.

In order to achieve these goals, host countries followed different paths in developing the capability of state petroleum enterprises in the oil sector. One method was promulgation of laws, regulations, and policies forcing the foreign oil companies to train and develop nationals to assume important posts in the operating company. Another was participation of state petroleum enterprises in actual field work and operations. Such participation followed different types of cooperation and arrangements. Some enterprises followed the joint venture type of agreements with a variable percentage accruing to them (Algeria, Iran, and Saudi Arabia). The service contract agreements, with foreign oil companies to be the operators, were even more common (as in Iran, Iraq, Egypt, and Lybia). Other countries preferred production sharing agreements (Indonesia, Syria, and Egypt). Participation of state petroleum enterprises in field work and in procurement of new projects and objects in the 1960s and early 1970s not only developed the capabilities of the state petroleum enterprises, but even more importantly broke the psychological myth of superiority which had been established by the major oil companies.

48

Few developing countries were able to form their own state petroleum enterprises with the capability of executing field work and other operations independent of the foreign oil companies. In some phases of oil operations, the state petroleum enterprises themselves employed specialized firms to enable them to execute some of their operations without the help or intermediation of the foreign oil companies. Indeed, most of the oil projects were implemented by these specialized firms, and it seems that this trend will continue. The emergence of national oil companies, especially in the OPEC countries, is considered to be one of the outstanding developments in the oil industry.

The Iraq National Oil Company (INOC) was established in 1964. It remained, however, ineffective and inactive until the revolution of July 17, 1968. Prior to this, INOC's staff, which hardly exceeded 100, was mainly involved in concessionary matters and some limited supervisory work on a service contract. Immediately following the revolution, great efforts were made in spite of the difficult economic warfare launched by the oil cartel and their hostile governments. INOC has succeeded in implementing the national exploitation policy of the Iraqi government by developing the North Rumaila Oil Field as well as other fields, and by creating a technical and administrative staff capable of implementing this government policy in the exploration, drilling, development, production, transportation, and marketing stages of the crude oil industry of Iraq.

In the field of exploration, which had been suspended for more than 10 years previously, work started rapidly. At present, INOC has six seismic crews with modern equipment. One of these crews uses vibroseis, other special crews operate in marshes. INOC also undertakes planning and direct supervision of the operations of nine seismic contracted crews.

For interpreting seismic data an interpretation center was created. It is equipped with an advanced digital computer and is run solely by national personnel. This center will be expanded to keep pace with the projected expansion of seismic survey operations. There are also national working groups undertaking interpretation work for the national crews as well as follow-up of the interpretations of foreign teams. In order to raise the efficiency of the Iraqi crews to levels comparable with those of foreign crews, it was necessary to create workshops and engineering and technical backup for the said crews. These exploration activities have led to the discovery of structures which have increased the expectation of the country's oil reserves reaching high levels.

As for drilling operations, in the late 1960s, these were confined to a limited effort by INOC under the terms of its service contract with ELF-ERAP and to the running of a single drilling rig by Iraq Petroleum Company in the Kirkuk area in the northern part of the country. In order to effect the required expansion in drilling operations, INOC took a number of steps. These included fixing a target for the requirements for the intermediate term, preparing specialized personnel and developing an organizational structure for operation management, purchasing rigs and drilling equipment, and entering into contracts with foreign contractors to help complement the national effort.

The expansion of drilling operations carried out directly by INOC was considerable. At the beginning of 1976 there were 23 operating rigs owned by the company and 13 owned by several contractors. National personnel assumed various responsibilities with respect to contractor's rigs, including logging, cementing, acidizing, and well completion operations. During this period intensive efforts were made toward preparing the required national personnel to operate drilling equipment and rigs, geographical expansion of drilling operations to cover most parts of the country, an increase in INOC's commitments in regard to contractors' rigs, and a speedy expansion in engineering services to keep pace with the expansion of the drilling operations and area.

At present, 42 drilling and workover rigs are operating in the country; INOC owns half of them. Iraqi personnel plan and supervise all drilling operations and operate about half of these rigs. In addition to the continuous expansion of drilling operations, INOC was able to improve the efficiency of rigs utilization and worker performance in all activities contributing to drilling operations. To give an idea of the expansion in drilling operations, it suffices to say that the volume of work has increased eight-fold over the period from 1972 to 1977.

Production operations entrusted to the national oil organizations also underwent a radical change from 1971 to 1977. In 1971, INOC's production was less than one percent of Iraq's total production of 82 million tons for that year. The year 1972, however, was a memorable one. During that year direct national exploitation was realized through the inauguration of the first stage of the North Rumaila Field. This was followed during the same year by the nationalization of Iraq Petroleum Company's operations. As a result, the share of the national sector in the total production rose to 30 percent. Consequent to successive nationalization acts and to the stepping up of production from the fields exploited nationally, the state petroleum enterprise was able to produce around 120 million tons during 1976. This accounted for Iraq's total production for that year, with the exception of 0.25 million tons produced under ELF-ERAP's service contract.

During 1977, INOC achieved full control over national production operations. Nationals now operate all production, treatment, and transportation installations. Production planning through follow-up reservoir behavior and periodic field measurement are also carried out by nationals. Recently, comprehensive long- and medium-term plans for oil exploration and production were prepared to consolidate the principle of integration of the oil sector within the country's economic plans. Planning for field development is carried out on a scientific basis and in accord with INOC's goals.

Associated gas is utilized in industrial projects as much as possible. This is to avoid flaring it, pending completion of major projects for exploitation of all associated gas.

As for INOC's major projects, many of these were executed during the last six years. The most important are the following:

- The North Rumaila Field development. This was executed in three stages: the first with a production capacity of 5 million tons per year, the second with a capacity of 18 million tons per year, and the third with a capacity of about 40 million tons per year. Another stage was added to put Mishrif Reservoir into production.

- Development of the Nahr Umr Field.

- Development of the Luhais Field.

- Development of the Jambur and Bai Hassan Fields.

- Construction of a new process plant near Kirkuk.

- The expansion of production and export capacity in northern Iraq.

- Development of the Buzurgan and Abu Ghirab Fields.

- Expansion of production and export capacity in the Rumaila and Zubair Fields.

In addition to the construction and expansion of Iraq's main pipelines undertaken as part of the oil field development, two major pipelines were completed during the period 1973 to 1977. The first is the Strategic Pipeline which links the northern and southern pipeline networks. Pumping is possible in two directions, thereby providing flexibility in the export system. The other is the Iraqi-Turkish Pipeline, which transports production from the Kirkuk area to Dortoyl on the Mediterranean coast of Turkey. The construction of the portion of this pipeline within Iraq was solely the responsibility of INOC. The giant Al-Bakr Terminal has also been completed. It consists of an island capable of loading tankers of up to 350,000 tons. This is one of the biggest and most modern terminals in the world. The first stage of the South Gas Project has also been completed. It provides for natural gas gathering from different stages of the various degassing stations. The second stage will include compression and liquefaction phases.

Implementation of water injection in the Rumaila Field is a large project designed for pressure maintenance and secondary recovery. The project is being executed at present, and its first stage is expected to be inaugurated soon. Smaller projects required for production and drilling are carried out either directly by INOC or through local contractors. Major projects are implemented by the State Establishment for Oil Projects through foreign contractors. The planning for such projects is carried out by the Establishment's national staff. It is noteworthy that the civil engineering works for most of the major projects are performed by the members of the Iraqi Construction Contracting Company.

In the field of crude oil transportation, INOC now possesses 15 tankers. Of these, seven are of 35,000 dwt, and the rest are of an average capacity of 150,000 dwt each. Most of the crews are still foreign. Efforts are being made to replace them gradually with Iraqi technicians and crews. Intensive training programs were started in 1972 with a view to preparing Iraqi officers, engineers, and technicians and qualifying them for assuming responsibility on board Iraqi tankers. INOC has also participated in the joint Arab efforts in the area of shipbuilding, repair, and maritime studies, as well as in the Conference of Arab Hydrocarbon Transporting Companies.

The State Establishment for Oil Refining and Gas Processing was established in 1972 to undertake the operation and maintenance of all governmental refineries, as well as to meet the country's requirements for petroleum products. The Establishment now operates the following refineries: Daura, Basrah, Kirkuk, Qaiyara, Alwand, Haditha, and Samawa. The total capacity of these refineries amounts to 8.5 million tons per year. The Establishment also operates a sulphur recovery plant near Kirkuk and a petroleum gas plant near Baghdad. The petroleum equipment plant is also attached to this Establishment.

The Establishment concentrates on operation and maintenance of the refineries by national personnel. Moreover, the Establishment utilizes its own personnel on a number of projects intended to increase the refining capacity or improve the quality of petroleum products. This has led to a substantial decrease in construction cost in certain cases and of training nationals to assume greater responsibility. An example of such a project is the undertaking by nationals of the mechanical erection work that doubled the capacity of the Basrah Refinery. This Establishment also conducts studies concerning local market demand for petroleum products and potential modification of the refining process to produce new products or to upgrade quality. It also investigates possible use of inferior quality crudes in local refineries.

The State Establishment for Oil Products and Gas Distribution undertakes distribution of oil products and gas to consumers throughout the country. Its functions include balancing production fluctuations due to seasonal changes through transportation of surplus stock to special storage tanks. The Establishment also handles transportation of imported products. During the past eight years, demand for fuel has doubled, while demand for lubricants has tripled, and the Establishment has succeeded in meeting this increase in demand within the country. The Establishment also provides airplanes with fuel and distributes LPG through carrier, railways, and pipeline. Transport vehicles increased eight-fold within ten years, and at present number around 2,000.

There has also been an obvious shift towards using pipeline transportation in recent years. In 1977, pumping started through the Baghdad-Basrah Products Pipeline. This pipeline is capable of pumping several products in two directions, and this is considered the most important project for product transportation. Construction of similar pipelines to provide flexibility in the distribution system has also been planned. Several other pipelines exist for transport of specific products.

The State Establishment for Oil Products was established in 1964 for design and construction of small petroleum projects such as service stations, reservoirs, storage tanks, and the like. After 1968, the Establishment's role underwent a radical change towards active participation with INOC and the National Iraqi Minerals Company in the implementation of national exploitation of oil and sulphur. This was represented by two pioneering projects: exploitation of Mishraq sulphur, and development of the North Rumaila Field. The Establishment's role consisted of technical and design work for these two projects as well as supervision of their implementation.

By 1972 the number of major oil projects had increased, and the Establishment began to deal with major oil and mineral projects, becoming the engineering center of the oil sector. It was entrusted with engineering consultation, feasibility studies, and design and supervisory work for major oil projects. In that period these projects were executed by foreign contracting companies.

By the end of 1975, however, a new phase began. The Establishment was entrusted with actual construction, and it became an independent national body responsible for construction of oil projects. The Establishment is now responsible for design and construction, including supervision of work by foreign contractors.

Efforts are being made at present to increase the Establishment's capacities through purchasing required equipment and training personnel for independent operation. A number of major projects have been completed under the supervision of the Establishment during the last five years.

In 1972, 21 projects were completed, including the first stage of development of the North Rumaila Field and Mishraq sulphur. The second stage of North Rumaila development and the Basrah Refinery construction were also started. In 1973, 14 projects were completed, in addition to the continuation of work in the second stage of the Rumaila project and the Basrah Refinery. Work began on the project of Nahr Umr development as well as on the Strategic Pipeline and the Al-Bakr Terminal.

In 1974, four projects were completed apart from the completion of the Basrah Refinery and the second stage of Rumaila. Work continued on the third stage of Rumaila, the Al-Bakr Terminal, and the Strategic Pipeline which were completed in 1975. Furthermore, the project of Nahr Umr development was commissioned. About 45 projects were supervised by the Establishment in 1976 including the Somalia Refinery project at Mogadishu, Somalia. The Establishment's activities were also diversified in 1977 with the commencement of direct implementation works. During the next several years, the Establishment will be responsible for three large projects: the North Gas projects, the South Gas project, and the Baiji Refinery.

The petroleum industry has witnessed prosperous periods for over a century. Discoveries have been far in excess of consumption, and oil has been available at very low prices. Previously, major oil companies dominated the scene unchallenged, while the governments of developing countries were helpless. Today, there are more than 60 fully owned or controlled state petroleum companies in the world. Historically, the petroleum industry has depended on a core technical staff and advanced technology, and it is anticipated that the role played by these factors will be fundamental to the future of the industry.

At the time of transfer of power and activities from the major oil companies to the state petroleum enterprises, while shouldering their newly acquired responsibilities, national staffs were faced with greater problems. In addition to running the transferred operations, they have to resolve problems inherited from the companies and, at the same time, bear any additional duties assigned to them by the state petroleum enterprise and those arising from shortages of trained staff. As new oil reserves become harder to find and more difficult to produce, the value of petroleum will become so high that governments should not plan to remain simple producers, but they should concentrate on developing downstream operations. Therefore, national staffs should develop their technical skills and know-how in anticipation of the accompanying greater responsibilities.

Exploration is a very important phase of petroleum activities, one which calls for advanced and sophisticated technology. Recently, many state petroleum enterprises have carried out exploration operations successfully and efficiently. INOC experience is not unique among developing countries. Indeed, SONATRACH of Algeria, Petrobras of Brazil, PEMEX of Mexico, ONGC of India, and others have had similar experiences. These enterprises expanded the area of their activities beyond national borders to other developing countries and have obtained good results. This success has been apparent in operation management, development of indigenous seismic crews and processing centers, and in attainment of a high standard of competence in solving complex problems.

The major oil companies generally concentrated on selectively exploring easily accessible areas with good potential. They showed interest only in large structures. On the other hand, the state petroleum enterprises have had to search for oil in more hostile environments and look for smaller structures or stratigraphic traps.

In the late 1960s, the average rate of discovery of oil in the world was about 25 billion barrels annually. This declined by 1975 to less than 16 billion barrels, of which 25 percent was credited to national oil companies and government enterprises. There has also been an increasing trend in the percentage of oil discovered by state petroleum enterprises. Recently published data on exploration activities, excluding the USSR, China, and eastern Europe, shows that during the period from 1970 to 1973 seismic surveys maintained a static level of around 7,000 crew-months annually. Western Europe and the developing countries accounted for 1,200 crew months. Of the latter, 30 percent were owned or contracted by government enterprises in 1973. The published data also indicate that seismic exploration activities reached a peak within a year after the crisis of 1973, and have decreased since late 1974.

During this same period, exploratory drilling maintained a steady level of around 2,000 wells annually outside the United States and Canada. State petroleum enterprises accounted for 50 percent of this drilling. Most experts expect that consumption of petroleum will soon bypass discovery. However, Mr. Halbouty claimed last year that almost half of the world reserves are yet to be discovered.

A review of published literature on world reserves was undertaken to develop conclusions on levels of world reserves. This, however, was difficult because these figures rose considerably every time the majors faced difficulties with host governments. The best example is the drastic change of world reserves before and after the 1973 crisis. Figures quoted before the crisis ranged between 350 to 450 billion barrels of proven recoverable world oil reserves. After the crisis the figures started climbing very quickly till they reached some 600 to 700 billion barrels. With the absence of sizeable new discoveries added or announced, it remains to be seen whether this is one of the myths of the informative major oil companies, or whether it is part of their doubletalk to producers and consumers.

Looking into the future, it may be expected that there will be a consolidation of the role of state petroleum enterprises in exploration and discovery activities. It is likely that more success will be met in discovering gas than oil. Adding new oil reserves will become increasingly difficult, calling for more improved techniques than are available today. Geologists will have to learn more through continuous study and research on sedimentary basins and their ecology. Geophysicists have to undertake more detailed seismic surveys, utilizing the most advanced technology, to be able to locate and define small and complicated structures, stratigraphic traps, lithology, bright spots, and the like. Seismic parties should be modified to enable them to undertake work in difficult areas. Alternative sources of seismic energy for different terrain should be sought and used on a wide scale. To achieve higher resolution of such data, large sophisticated computers should be developed.

Exploratory and development drilling is aimed for deeper horizons, spreading to deeper offshore locations and more difficult terrain, and facing stricter environmental, conservation and safety regulations. The technical personnel have to develop their capabilities and skills to cope with these factors. They have to overcome problems of higher pressure and temperature, and must develop familiarity with mud control problems and the selection of proper equipment for wells to achieve their objectives economically. Down-hole services are still highly monopolized, and unceasing efforts have to be made to master these by national staff.

Turning to production, policies adopted previously by the major oil companies sought the highest profit irrespective of long-term recovery and field conservation. As a result, most of the fields were quickly depleted or abandoned at critical stages because of profit considerations. National staff have an important role to play in design of optimal production policies to obtain the greatest recovery from existing fields. National staffs are again called upon to study pressure maintenance operations in conjunction with enhanced recovery techniques to achieve maximum recovery factors. Such activities are now being undertaken in a number of developing countries including Algeria, Brazil, Chile, Colombia, Libya, Mexico, Trinidad and Tobago, Venezuela, and others.

The anticipated decline in production from classical fields will undoubtedly lead to the increased usage of more viscous and sulphurous crudes. Increasing recovery from such fields will require tremendous efforts by researchers. Processes like steam soak, in situ combustion, and steam injection are already gaining in popularity. In Venezuela alone there are more than 70 operating thermal projects. Miscible drive projects have also been applied in the United States and Canada. Some developing countries like Algeria and Libya have started such projects to improve recovery. Viscous water (obtained by addition of polymers, for example) is being experimented with in many United States and Canadian reservoirs.

Technical staff will have to deal with new production facilities, some of which are highly automated, remotely controlled, and suitable for difficult producing conditions. Corrosion problems will become serious as installations age and as the utilization of chemicals increases. Attention has to be given to overcoming these problems through research. Well servicing, workovers, subsea completions, and offshore platforms will increase with time, and personnel should have the training to handle them.

In considering pipeline technology, a number of developing countries have achieved limited capabilities in laying small-size pipelines for oil and gas. It is expected that future demand for oil and gas will call for an expansion of pipeline networks as a result of the following factors:

1. increased utilization of natural gas by producing countries through the prevention of gas flaring;

2. increase in demand for oil stemming from the growth in energy consumption;

3. increase in the producing countries' participation in downstream operations; and

4. reliance on pipeline product transportation systems in preference to other modes of distribution.

These developments will also call for improvement in the technology of pipeline manufacture and in the techniques of welding. During the next decade, many producing countries will concentrate on the technology of pipeline manufacture and construction as a means of upgrading their industrial capabilities. This aspect could be considered as a short-term objective of state petroleum enterprises. Pipeline network complexity is expected to call for improvement in automatic control systems, with the accompanying technology for operation and maintenance of such systems.

The process of crude oil refining has witnessed major advances in the diversification of products and improvement of product quality. Developing countries have never been allowed to expand refining capacities to levels beyond their internal consumption, except in cases where this has been advantageous to the major oil companies. It is our view that in the future, petroleum refining will meet other challenges and changes. The present drive in the industrialized countries to limit crude oil utilization and to maintain demand for oil at present-day levels will result in the restriction of oil to petrochemicals and the gradual elimination of its energy role. This will require that refining processes be modified to provide for this shift. In the past, companies have concentrated on producing better quality crudes, thus leaving processing of heavier crude oils with higher sulphur content largely undeveloped. The ongoing depletion of world oil reserves will necessitate the development of refining technology to treat these crudes in ways acceptable to environmental conservation laws.

The participation of producing countries in this phase of the industry is inevitable. It is an internal part of their drive to attain new capabilities in downstream operations through construction of new refineries. The present refining capacity in the industrialized world is not fully utilized, because it was built to meet consumption patterns based on higher growth rates of energy demand. This excess capacity will prove short of future expected demand, and the need for further capacity will be evident. Refining in oil producing and developing countries has clear advantages. However, the present distribution of up stream operations in oil-producing countries and downstream operations in oil-consuming industrialized countries can become the means of technology transfer from the industrialized countries to developing countries. Automatic control of refinery operations to reduce the requirements in manpower and to ensure more efficient operation is a natural outcome of the development in refining technology.

Producing countries have realized the importance of natural gas, both as a fuel and as an essential feedstock for petrochemical complexes. The present drive towards conservation of associated gas through utilization in industrial projects, petrochemical plants, or reinjection in oil reservoirs is expected to lead to the establishment of large gas processing and petrochemical complexes in these areas.

The evolution of the petrochemical industry in the industrial countries during the last quarter of the century has been very impressive. The tremendous success in manufacturing synthetic materials has been due to the availability of low-cost petroleum feedstocks. The petrochemical evolution was achieved by maintaining a continuous balance between cost, technology, and demand. However, the versatility of petroleum as raw material and its flexibility in refining processes was the main reason for this transformation. This is the aspect of petroleum utilization that is the most stable in the medium term. Therefore, it is suggested that petroleum enterprises in developing countries undertake such projects in the near future. This could be a preliminary step to reducing the costs of transportation and manufacture of these goods.

A few national petroleum companies in producing countries have acquired oil tankers. This trend has been discouraged during the past few years by the low oil transportation costs prevailing in the world as a result of the oversupply of crude in the market. However, this should not distract form the value of this aspect of the industry since it is complementary to the principle of sovereignty of producing countries over their natural resources. It also offers flexibility in producing crude or products and for different patterns of internal industrialization. Need will arise in future for transporting LPG and LNG to foreign markets, and the various national enterprises should take steps towards securing the utilization of appropriate tankers in the future.

II

Cooperation Among State Petroleum Enterprises

7 Cooperation Among State Petroleum Enterprises

M. Nezam-Mafi

Starting more than 50 years ago, the oil business in most parts of the world (except for the socialist countries, the United States, and Canada) was under the control of the seven major oil companies. However, political conditions changed with World War II, when some of the former colonies of the United Kingdom, France, and The Netherlands obtained their freedom. Concurrently, there were basic changes in patterns of energy consumption as the share of oil and gas in the total world energy picture increased. With the advancement of technology, most of the newly independent countries established their own state petroleum enterprises, or strengthened the enterprises already in existence.

These state petroleum enterprises were established in order to satisfy requirements for internal consumption, ensure proper exploitation of national resources, decrease dependence on foreign oil companies, and obtain a share of the global oil business. But the level of activity any state petroleum enterprise could attain was dependent on the petroleum reserves and level of technology of each. Accordingly, various types of government companies were created, depending on the different conditions in each country.

The most important producing country enterprises are the members of the Organization of Petroleum Exporting Countries (OPEC), most of which were established in the 1950s and 1960s. The oldest among them is the National Iranian Oil Company (NIOC), which was established in 1951, after the nationalization of the oil industry in Iran. Later, Eduador, Iraq, Libya, Nigeria, Qatar, Saudi Arabia, United Arab Emirates (Abu Dhabi), and Venezuela established their own state enterprises. In addition to the OPEC member countries, other developing countries with exploitable oil resources, especially in Latin America, also established their own state oil enterprises. In recent years, with the discovery of petroleum in the North Sea, industrial countries such as the United Kingdom and Norway have founded state petroleum enterprises as well. It is anticipated that in the next few years they may have considerable oil for export.

In the oil-importing developing countries, such as India and Turkey, state petroleum enterprises have been created to provide petroleum products at low prices for internal consumption. These companies have tried to reduce national dependence on transnational oil companies and to control domestic distribution of oil products. They have various special rights in their home countries, and often use the legislative power of their governments to assist them in obtaining secure supplies at the best available terms.

An interesting example of such use of government power, in this case by an industrialized country, was the Sahara Petroleum Law passed by the French in 1958 with the objective of guaranteeing that at least part of French internal requirements could be met from sources under French control and by French companies.

After World War II, as a consequence of increased oil requirements and the strategic importance of petroleum supply, the industrialized countries began to question whether they should continue to depend on the transnational companies for domestic supply and exploitation of reserves as they had in the past, or whether they themselves should establish state petroleum enterprises.

Also coming into play at that time was the elevated income level of the national petroleum companies, which was largely due to the profitability of their operations in the Middle East. Because of the low production costs (in 1945 the cost of producing one barrel of oil in Saudi Arabia was 19 cents and 10 cents in Bahrain (1), the small payments made to the producing countries (about 25 percent of the total value of crude sold, under the financial terms of Iran's 1933-1951 agreement), and the large profit margins (even after the 50-50 concession agreements), from 1956 to 1960 the net profits of these companies averaged 66 percent from their operations in the Middle East. (2) These profits alone were sufficient encouragement to the industrialized countries to form state petroleum enterprises.

However, because state petroleum enterprises in the industrialized countries had the advantage of advanced technology, they did not limit their activity to securing adequate supplies for domestic consumption. (3) They began to try to obtain a share of the international oil business.

The transnational oil companies were very unwilling to let other companies have any part of the world oil business. In the Middle East, especially, they had operated as an effective cartel since the conclusion of the Achnacarry and Red Line Agreements of 1928. The transnational companies had detailed agreements on the division of markets and Middle East concessions, and used them to prevent new arrivals from entering the market. (4) As a result, as recently as 1953 none of the industrialized country state petroleum enterprises had substantial crude oil production under its control. In view of this situation, the state enterprises proposed a number of new agreements to the producing countries that would permit them to participate in exploration, production, and marketing. These agreements, in the form of joint ventures and service contracts, marked the beginning of cooperation between the national oil companies of the producer countries and the state petroleum enterprises of the industrialized countries.

The Italian state petroleum enterprise, Ente Nazionale Idrocarburi (ENI), which was established after the discovery of natural gas in the Po valley in 1953 by a merger of Azienda Nazionale Idrogenazione Combustibili Generale Italiana Petroli (AGIP) and Societi Nazionale Metanodothi (SNAM), pioneered the joint venture contract in the petroleum industry. The joint venture was developed by ENI not only gave the producing country better terms than those offered in the transnationals' 50-50 concessionary agreements, but the ENI contracts also stipulated that in the event of a discovery, the producing country could participate in the production operations. This was the first instance of such an opportunity. Under these new contractual agreements, the three elements which were lacking in the oil producing countries - financing, know-how, and markets - were supplied by the foreign partner.

It was the French who pioneered the service contract through the Entreprise de Recherche et d'Activite Petroliere (ERAP), which was formed in 1966 by the merger of the Bureau de Recherche Petroliere (BRP) and Regie Autonome des Petroles (RAP). Like ENI, ERAP offered a new type of agreement to the producer countries in order to gain a foothold in the Middle East. Under the service contract, ERAP provided financial and technical services as a general contractor for the national petroleum company and was remunerated through guaranteed sales of a certain amount of oil.

Both the joint venture agreements and the service contracts resulted in greater income for the oil producing countries as well as more control of their petroleum resources. The industrialized countries gained the benefit of a secure source of supply, and their petroleum companies acquired new opportunities to play a more active role in the international oil business.

Following the ENI and ERAP agreements, the state petroleum enterprises of Japan, the Federal Republic of Germany, Austria, and Spain, as well as a number of other government organizations, concluded joint ventures or service contracts with the producer countries (see table 7.1.).

Most of the cooperation agreements between state petroleum enterprises that have been made so far fall into one or another of several main categories: between state petroleum enterprises of producer and industrialized countries; between state petroleum enterprises of producer countries; between state petroleum enterprises of producer countries and either developing countries or the state agencies of socialist countries; and those between the Latin American state petroleum enterprises.

Cooperation between the state petroleum enterprises of producing countries and those of industrialized countries has its roots in the desire of the industrialized countries to gain access to the oil of producer countries. This cooperation provided substantial benefits to the producing countries because the agreements enabled their state enterprises to take part in production activities and also provided them with an opportunity to sell part of their production directly in the market. This, in turn, created a measure of competition for the

TABLE 7.1. State Petroleum Entities in the Middle East

State Entities	Iran	Iraq	Kuwait	Neutral Zone	Qatar	Saudi Arabia	United Arab Emirates
Abu Dhabi National Co.	-	-	-	-	-	-	P
ELF-ERAP (France)	P	X	-	-	-	X	-
Deminex (F.R. of Germany)	X	-	-	-	-	-	-
ENI (Italy)	P	-	-	-	P	-	X
Hispanoil (Spain)	X	-	X	-	-	-	P
Indian Government	P	X	-	-	-	-	-
Iraq Naitonal Oil Co.	-	P	-	-	-	-	-
Japan Petroleum Development Corporation	X	X	-	-	X	-	P
Kuwait Government	-	-	P	P	-	-	-
Kuwait National Petroleum Co.	-	-	X	-	-	-	-
NIOC (Iran)	P	-	-	-	-	-	-
OMV (Austria)	X	-	-	-	-	-	-
Pakistan Government	-	-	-	-	-	X	-
Petrobras (Brazil)	-	X	-	-	-	-	-
Qatar Government	-	-	-	-	P	-	-
Saudi Government	-	-	-	P	-	P	-

P = Production
X = Exploration

transnational oil companies in the international market. In some cases, as for example the agreement of Japan Petroleum Development Corporation (JPDC) with Saudi Arabia and Kuwait concerning the Neutral Zone, provision was made for limited partnership in downstream operations.

It was hoped at the time, and specifically mentioned in some of these agreements (for example the one between NIOC and ERAP), that the cooperation in oil production would lead to large-scale economic cooperation between the respective countries in other fields. However, in practice, this hope for the most part went unrealized, and the agreements remained limited to petroleum cooperation. An agreement combining oil cooperation with general provisions for the economic development of the producer country was concluded between Algeria and France in 1965, three years after Algeria gained independence. Under this agreement, the Algerian national oil company (SONA-TRACH) and a subsidiary of ERAP established a cooperative association, which was charged with carrying out exploration and exploitation activities in a vast area of Algeria, while at the same time envisaging an organization for industrial cooperation for the economic development of Algeria. This agreement encountered difficulties from the beginning. In 1971 it was effectively discontinued due to the nationalization of up to 51 percent of the shares of the oil companies operating in Algeria.

The history of the development of the transnational oil companies shows that cooperation among them increased their power and capabilities. In their efforts to exploit their petroleum resources and to penetrate markets, the state petroleum enterprises of producer countries were faced with competition from the transnational oil companies and, in these circumstances, it was natural that the state petroleum enterprises should wish to close ranks.

Cooperation among state petroleum enterprises of producer countries began at about the same time that agreements were made with the state petroleum enterprises of the industrialized countries. Producer country cooperation has been effected in two ways: by direct cooperation, and by cooperation through OPEC and OAPEC.

The establishment of direct cooperation between state petroleum enterprises of producer countries was a logical step, because most of these enterprises were established in petroleum producing countries with broadly similar economic situations and, as production increased, all became increasingly dependent on oil income. None of the producer countries could forego this income without experiencing profound economic disturbances. Furthermore, all were hindered in the exploitation of their resources by shortages of specialized manpower, lack of technical know-how, inadequacy of information about the oil business, lack of markets and technology, and insufficiency of capital. As a result, whenever these countries attempted to take over oil operation in their countries (as happened, for example, in Iran in 1951), they faced major problems. For these reasons, the efforts of the producer country enterprises in the first few years of their existence were generally limited to the fulfillment of their responsibility to handle internal distribution.

Cooperation among the enterprises began with the commencement of regular exchanges of information in the early 1950s, following the conclusion of the 50-50 concession agreements. The major reason for these exchanges of information was the inclusion of "most favored nation" clauses in these agreements. As a result of these contacts, the state petroleum enterprises became aware that each of them had different potentials in the various fields relating to oil exploitation, and they realized that there were possibilities of fruitful cooperation between them.

It appears that the cooperative efforts of the National Iranian Oil Company have been the most varied. One of NIOC's important cooperation agreements was with the state petroleum enterprise of Libya. This was first implemented in 1965, with the dispatch of several specialists, and it continued for several years. NIOC also cooperated with Algeria's SONATRACH on the basis of a technical cooperation agreement between the two enterprises; and from 1969 to 1974, several groups of NIOC's specialists worked in Algeria, both in the central offices of SONATRACH and in the southern oil fields.

Of all the agreements, it is probably those establishing cooperation through OPEC that have had the greatest effects. The most important aims of OPEC, as set out in its first resolution, consisted of maintaining oil prices and regulating production of OPEC member countries. The second OPEC resolution proclaimed that uniformity of petroleum policy among the member countries was the best means of safeguarding their interests. (5) It was obvious that close cooperation among the state petroleum enterprises would be necessary in order to give effect to the resolutions.

The first meeting of representatives of state petroleum enterprises arranged by OPEC was held in 1966 in Caracas. It was here that the representatives of OPEC state petroleum enterprises decided to establish a committee to study possibilities of cooperation among state petroleum enterprises. This committee held meetings until 1970, and made proposals concerning coordination of state petroleum enterprises' petroleum policies with regard to the international petroleum business, exchange of technical information, and elimination of competition among producer countries.

After the oil price increase in 1973 and the resulting improvement in the financial situation of the OPEC countries, together with the takeover of petroleum industries by the producing countries, cooperation among the state petroleum enterprises was highlighted in the member countries.

Representatives of the OPEC members' state petroleum enterprises met in London in August 1974, and the basic aims of cooperation were agreed upon as follows:

1. to exchange information relating to price trends and rates of production;

2. to exchange relevant information on new deals concluded during the preceding period; and

3. to coordinate marketing systems and contracting conditions of member's state petroleum enterprises.

No immediate practical action was taken to implement these aims. However, at the ministerial conference of OPEC held in June 1975 in Gabon, the following resolution was adopted:

> To take the necessary measures for the promotion of cooperation among the state petroleum enterprises of OPEC member countries, particularly in the field of marketing, and in this context, decided to create such organs and institutions within the framework of OPEC.

Following the OPEC conference in Gabon, one of the OPEC efforts to effect cooperation between state petroleum enterprises was the organization of a seminar on "The Present and Future Role of the National Oil Companies," which was held in October 1977 in Vienna. The participants in the seminar, besides OPEC state petroleum enterprises, included 43 delegations from other oil companies, including the transnationals. After the failure of the North-South Conference in Paris, this seminar provided a good opportunity for different groups of oil companies to express their views on problems and possibilities of cooperation.

In 1968 three countries - Libya, Kuwait, and Saudi Arabia - were instrumental in establishing the Organization of Arab Petroleum Exporting Countries (OAPEC). Seven other Arab countries - Algeria, United Arab Emirates, Qatar, Bahrain, Iraq, Syria, and Egypt - joined the organization. Article 2 of the OAPEC agreement defines its main purpose as follows:

> The principal objectives of the organization are the cooperation of the members in various forms of economic activity in the petroleum industry, the realization of the closest ties among them in this field, and the determination of ways and means of safeguarding the legitimate interests of its members individually and collectively; the unification of efforts to ensure the flows of petroleum to its consumer markets on equitable and reasonable terms, and the creation of a suitable climate for capital and expertise invested in the petroleum industry in the member countries.

Of particular importance in recent years has been cooperation between the state petroleum enterprises of producer countries and the developing countries. Although the joint ventures and service contracts put a part of crude oil production at the disposal of the state petroleum enterprises, the markets of the technologically-advanced, industrialized countries were already controlled by transnational oil companies. Therefore, the producer countries turned to the developing countries for their marketing, and the state petroleum enterprises took steps to market their oil and to establish partnerships in the other activities pertaining to oil. In 1963, NIOC began marketing its oil directly and was followed later by the state petroleum enterprises of Algeria, Libya, Nigeria, and Indonesia.

NIOC has steadily increased the quantities of oil it markets directly, as a result of greater available quantities through its agreements with IPAC, 1963; LAPCA, 1968; IMINOCO, 1971; and the take-over of the Iranian Oil Consortium. Direct marketing totaled 1,520,000 barrels per day (b/d) during the first ten months of 1977. Since the time it began marketing oil directly, NIOC has entered into marketing agreements with countries or state oil enterprises in such developing countries as Bangladesh, Pakistan, India, Sri Lanka, The Philippines, Zaire, Tanzania, Senegal, Egypt, Morocco, Brazil, Argentina, and Chile. NIOC has also participated in a number of joint ventures abroad, including a refinery and fertilizer plant in Madras; refineries in South Africa, South Korea, and Senegal; and an oil products distribution undertaking, also in Senegal. The agreement to set up the Madras Oil Refinery was signed by NIOC, Pan American Petroleum Corporation, and the government of India. The government of India owns 74 percent of the shares with the remaining 26 percent divided equally between NIOC and Pan America. The refinery has been in operation for 11 years with crude supplied from Iran's southern oil fields. A fertilizer plant situated next to the refinery has an annual capacity of 247,500 tons of ammonia, 292,000 tons of urea, and 544,000 tons of NPK fertilizer.

The South African refinery was established through an agreement between NIOC and the firm of National Petroleum Refiners of South Africa, which established the Sasolbourg Refinery as an independent industrial unit. NIOC has 17.5 percent of the shares of the undertaking. Other partners in this venture are the South African Coal, Oil and Gas Corporation with 52.5 percent, and Total Refining South Africa with 30 percent. The capacity of this refinery was 3.46 million tons in 1976. Crude is also supplied from the southern oil fields.

After an agreement with the Republic of Korea, a contract was signed between NIOC and Sun Yan Co. in 1975 establishing the Iran-Korea Company to build a refinery with a capacity of 60,000 b/d. This refinery was scheduled to use crude from southern Iran.

In partnership with Senegal's state petroleum enterprise (SOSERAP), NIOC established IRASENCO to build an oil refinery with a capacity of 30,000 b/d, develop an oil distribution network, and undertake the exploitation of mines. The first step taken by IRASENCO, in 1976, was to enter into a partnership with Shell Oil Company for oil distribution activities in Senegal.

With the increase in the consumption of oil products in socialist countries, and concurrent difficulties faced by the USSR in meeting the needs of these countries, barter contracts between socialist and oil-producing countries began to show an upward trend. A number of oil-producing countries have signed contracts for the sale of oil against the purchase of other goods from the socialist countries. In 1976, OPEC member countries (with the exception of Iraq, for which information is not available) exported an average of 123,400 b/d of crude to the socialist countries. (6) During the past few years, NIOC has signed contracts for the sale of oil against the purchase of other goods with such socialist countries as Rumania, Bulgaria, Hungary, Poland, Czechoslovakia, and Yugoslavia.

The most important contract was signed in 1966 with the USSR for the sale of gas. According to the Iranian gas line project, 10 billion cubic meters is exported to the USSR annually. The implementation of this project is the responsibility of the National Iranian Gas Company (NIGC) which, until recently, was a subsidiary of the NIOC and the Naftkhim Promexport of the USSR. A separate gas sales contract was signed between NIOC and the Soyuz Naftexport of the USSR. In exchange for the gas, the USSR will provide Iran with heavy machinery and equipment. The gas pipelines built for these sales were inaugurated in 1970.

Successful regional cooperation among oil exporting and importing developing countries is exemplified by the cooperation between state petroleum enterprises in Latin America. In 1964, an organization called "Asistencia Reciproca Petrolera Estatal Latinoamericana" (ARPEL) was formed by the oil enterprises of Uruguay, Ecuador, Venezuela, Chile, Colombia, Brazil, Peru, Argentina, and Bolivia. Other Latin American governments or state petroleum enterprises send observers to the meetings of this organization. The aims of ARPEL include the development of technical cooperation and exchange of know-how between the members; research on measures required to achieve greater cooperation between the state petroleum enterprises; implementation of activities to spread the oil industry in Latin America, and to develop industries producing equipment necessary for the oil industry; security of oil supply and harmonization of development plans; and convening of technical and educational congresses, conferences, and meetings to study problems of the oil industry.

Since 1966, a number of contracts and agreements have been concluded between the member organizations of ARPEL in the fields of pipeline oil transport, oil purchasing, research, establishment of petrochemical complexes, marketing, and transport of LNG and LPG, exploration and production, marketing of by-products, establishment of refineries, common use of loading piers, technical cooperation in the field of the development of alternative sources of oil such as oil shale, and the training of employees.

The North-South Conference in Paris aroused hopes of resolving certain economic and international cooperation problems. But when it ended in June 1977, after almost two years of deliberations, no substantial results had been achieved. Some of the most important areas of discussion were those mooted in the Commission on Energy, such as the protection of the purchasing power of the price of oil exports, a relationship between this price and the costs of production of substitute sources of energy, and a relationship between the price of oil and the price-index of industrial goods. Had the North-South Conference resulted in agreements, future oil cooperation between the industrialized countries and the Third World would have been based on greater mutual confidence.

During recent months, a number of barter contracts have been concluded between oil-producing and oil-consuming countries, including an agreement between Iran and Brazil, according to which Brazil has committed itself to import 25 percent of her oil needs from Iran, and Iran has agreed to spend 30 percent of the income from these exports on imports from Brazil and investments in that country.

Another such contract is the recent agreement between Algeria, Italy, and Tunisia for the first gas pipeline from North Africa to Europe. According to this agreement, Algeria will export 12 billion cubic meters of gas annually to Italy, starting in 1981. Tunisia will receive 5.25 percent of the exported gas as a transit toll. ENI's Saipem and Snamprogetti are responsible for the 1500 mile pipeline. At the same time, another agreement was signed by the two countries concerning the construction by the Fiat group of a complex at Oran to produce 100,000 cars per year.

Recently, Japan predicted that by 1990 one-third of Japan's total oil demand should be obtained through state-to-state agreements or direct deals and barter arrangements between Japanese firms and the state petroleum enterprises of producing countries. It is felt that such contracts between governments or government oil organizations in all the three groups of countries (socialist, industrialized, and developing) will be on the increase in the future.

All the estimates regarding the future use of energy published over the past two years have been similar in certain respects. They have agreed on the continued dependence of the world on oil till the end of this century, the increase in demand for oil beyond its potential supply possibilities beginning with the mid-1980s, and the basic role of the developing oil-producing countries in meeting global oil demand. The estimates of energy consumption of the industrialized, socialist, and developing countries bear witness to the more rapid increase in oil and gas consumption of the developing countries. According to one estimate, should the developing countries register an average annual rate of growth of 6.9 percent over the years 1970 to 2000 to diminish the gap between the rich and the poor countries, their oil and gas consumption will show an average annual increase of 9.7 percent and 11 percent respectively. (7)

At present, oil and gas constitutes an appreciable share of the imports of oil-importing developing countries. During 1974, this share averaged 9.5 percent for poor developing countries and 11.6 percent for other developing countries (table 7.2). The increased dependence on oil and the accompanying financial problems make it imperative that these countries take steps to meet a proportion of their requirements through their indigenous resources. If these countries cannot do so, they will be forced to continue to import their oil and gas from the oil-producing countries.

The oil-producing countries have also realized that in the long run, additional crude oil production alone will not help their economic development. They must enhance the value of crude oil by creating refining complexes, transportation and distribution facilities, and petrochemical plants. Although OPEC produces over half of the world's oil, it commands a meager 6 percent of refining capacity and 3.2

TABLE 7.2. Shares of Net Petroleum Imports in Total Imports in 1974

(Percentage)

Petroleum Importing Developing Countries		Least Developed Countries	
Africa		Afghanistan	14.1*
Morocco	12.0	Bangladesh	7.8*
United Republic of Tanzania	15.5	Benin	5.0
Kenya	13.5*	Bhutan	--
Zambia	11.0	Botswana	--
Ghana	13.6	Burundi	6.2*
Sudan	5.8	Central African Republic	11.9*
Others	7.3	Chad	4.3*
Total	9.4	Democratic Yemen	20.7
		Ethiopia	11.8*
Asia and the Pacific		Gambia	4.7*
India	27.6	Guinea	29.6
Republic of Korea	13.1	Haiti	10.3
Philippines	18.8	Lao People's Democratic Republic	10.1
Thailand	15.7	Lesotho	--
Turkey	17.1	Malawi	9.2
Hong Kong	5.9	Maldives	--*
Others	4.5	Mali	6.2
Total	10.3	Nepal	8.0
		Niger	7.0
Latin America		Rwanda	8.5
Brazil	20.4*	Sikkim	--*
Mexico	4.7*	Somalia	5.0
Cuba	26.5	Sudan	5.8*
Argentina	10.3	Uganda	0.5
Jamaica	19.9	United Republic of Tanzania	15.5*
Chile	11.5	Upper Volta	3.4*
Others	10.3	Western Samoa	4.2*
Total	13.9	Yemen	5.9
		Total	9.5
All petroleum-importing developing countries	11.6		

Source: "Recent Energy Trends and Future Prospects," paper presented to the Committee on Natural Resources at its fifth session (E/C.7/70, April 11, 1977).

* Estimate

71

percent of petrochemical capacity. Furthermore, only 3 percent of OPEC oil is transported in tankers owned by member countries. (8)

From the point of view of domestic market penetration, some of the old problems of the OPEC member countries have been solved. Similarly, today the domestic exploitation of the oil industry and oil-pricing policies are controlled by the countries themselves. With the increase in oil prices of 1973 and 1974, resulting in a flow of financial resources on the order of $80 billion per year from the rich countries to the poor ones, investment problems have been solved so much so that at the end of 1976, the foreign exchange reserves of Saudi Arabia, Kuwait, United Arab Emirates, and Libya totaled $94 billion. However, such problems of OPEC member countries as lack of technical know-how, markets and, to a certain extent, human resources still remain unresolved. The state petroleum enterprises of industrialized countries are in possession of technological know-how and markets. However, they lack oil and, in certain cases, capital as well. Cooperation between state petroleum enterprises should be planned in such a manner as to assure definite advantages to the various groups. It is felt that a good framework for such cooperation between the three aforementioned groups of countries could be concentrated in exploration and drilling, increasing the output ratio of wells establishment of a research center by the transnationals, the exchange of information, and the training of experts.

The undiscovered oil potential in the developing countries is generally estimated to be over 40 percent of the world's total. Experts at the Ministry of Energy of the USSR have put this figure at 40.5 percent, and at the latest OPEC seminar Mr. Halbouty estimated it at 42 percent. Furthermore, research carried out by the United States Geological Survey has noted possibilities of oil discovery in seven developing countries with an oil resource potential of between 10,000 and 100,000 million barrels each, another 30 countries with a range of 1,000 to 10,000 million barrels, and 16 countries with a range of 100 to 1,000 million barrels. (9) This seems to be very much in favor of the developing countries which, instead of seeking loans and aid for oil imports, can create favorable conditions for investment. One way of doing this is through three-way partnerships, one side of the partnership being the developing country and the other two sides being the state petroleum enterprises of producing countries. Such partnerships will have to take into account the interests of the three parties concerned. Cooperation in the field of exploration and drilling between the government oil organizations and oil producing countries can also be further developed. Recently, SONATRACH signed a contract with the ENI subsidiary SAIPEM for SONATRACH's drilling operations in the Sahara.

The increase of oil prices has made the implementation of plans for secondary recovery from oil wells profitable. It is in the interest of the oil-producing countries to increase the lifetime and recovery factor of oil wells. Most oil-producing countries have drawn up plans for increasing the output ratio of oil wells and the injection of oil or gas. The transnational oil companies, with years of experience in this area, have developed new methods such as underground blasting and the injection of chemicals with water. (10) Iran, Venezuela, and Algeria are implementing huge secondary recovery plans. NIOC investment on secondary recovery was estimated by its managing director earlier this year at $2.5 billion. It is hoped that this investment will increase the recovery factor in Khouzestan oil fields by 5 to 10 percent. In view of the huge investments that the state petroleum enterprises will be making, an exchange of information on the results obtained in this field, convening of conferences, and on-the-spot inspection are suggested.

One of the basic problems of the national oil companies is lack of technological know-how. Service imports of OPEC member countries have jumped from $8.8 billion in 1974 to $17.4 billion in 1976 and, according to estimates given by Chase Manhattan Bank, this figure is expected to soar to $33.1 billion by 1980.

At present, a number of OPEC member countries (Algeria, Iran, Saudi Arabia, Indonesia, Iraq, Kuwait, Libya, Nigeria, and Venezuela together with Egypt, which is a member of OAPEC but not of OPEC) have petroleum institutes under their state enterprises or under the relevant ministries. (11) But in practice, no close cooperation exists between these institutes. It is suggested that steps should be taken to bring about closer cooperation between these institutions, and also that the establishment of a research center in an OPEC member country should constitute part of the working program of OPEC member national oil companies.

Most transnational oil companies possess more detailed information and knowledge than the state petroleum enterprises of developing countries. This is no doubt due to the span of their operations, as well as their practice of withholding of full information on all their activities. In the past, lack of knowledge of the various facets of the oil industry has, in a few cases, resulted in incorrect decisions by state petroleum enterprises.

In this respect, the Centre of the United Nations Commission on Transnational Corporations should prove to be of great help in strengthening the position of government oil organizations. The Centre has established an information bank on transnational companies and maintains a roster of experts. It is suggested that OPEC also set up an information bank, and that the state petroleum enterprises be permitted to use the information in its possession. This should help the government oil organizations in their decision making processes. Moreover, the convening of conferences or seminars by state petroleum enterprises should also help in an exchange of ideas between such organizations.

Even today, in some oil producing countries the industry is not fully controlled by the state enterprises because it does not have the number of experts required to run the industry. Even those enterprises that have achieved greater success in this respect do not possess all the required experience and expertise in such areas as exploration and development, or the new areas such as petrochemicals, LNG, and LPG. Past experience has shown that technological cooperation between state oil enterprises has proved very useful. The same holds true for mutual training programs on a regional basis between government oil organizations in the Middle East, North Africa, Latin America, and the Far East.

Refining is another area in which cooperation can be especially important. The oil-producing countries refine varying amounts of their crude locally. In the OPEC member countries in 1976 the ratio of refining to production ranged between a minimum of 0.8 percent in the United Arab Emirates and a maximum of 63.4 percent in Venezuela. For two other major oil producers, Saudi Arabia and Iran, this ratio stood at 8 percent and 13.3 percent respectively. In order to boost their exports of oil products, the oil producing countries have drawn up plans to open a number of new refineries. Over the next few years, OPEC member country investment in new refineries is expected to be $5.6 billion.

In the past, some three-way or two-way joint ventures to establish refineries have been completed successfully, while some others are still being implemented. These include such joint ventures as the Madras Refinery venture between NIOC, Pan American Oil Company, and the Government of India, or the joint venture between Iran and Senegal in setting up the Senegal Refinery, or between Abu Dhabi and Pakistan to establish a refinery in Pakistan.

Due to transportation problems, there is much less experience with international trade in gas than with oil. However, future trade estimates for gas show a rapid increase in trade for this energy source. It is expected that the total world trade in LNG will reach 45 billion cubic meters by 1980, and between 150 and 230 billion cubic meters by 1990.

The known OPEC reserves of natural gas vary from country to country. Some of the countries which are rich in these reserves have started exporting their gas. Algeria, which has been exporting LNG to Canvey Island since 1964, has greater experience in this field than other OPEC member countries, and has envisaged an investment of $17 billion in this field over the next few years. Indonesia is in the process of implementing a multi-billion dollar project for the export of LNG to Japan, while Nigeria and Qatar each have a $4 billion project for the export of LNG to the United States and Japan respectively.

A prerequisite of gas marketing is huge investment. The major part of such outlays goes to the establishment of LNG plants, erection of piers, and the purchase of special vessels for the transport of liquid gas. The possession of information, the knowledge of the experience of others, and the implementation of joint research programs should prove very useful for countries active in this field. It is felt that wide-ranging, two-way, or multiple cooperative efforts on a large scale between various groups of companies is feasible.

The events of the past few years have enhanced the possibilities for creation of common markets between the oil producing countries and their nonoil-producing neighbors. With the implementation of extensive economic plans in the oil producing countries, salaries and wages, and in consequence, the price of goods manufactured in these countries have shown an upward trend. Some of these countries have also started investing in heavy industries. Barring problems likely to spring from markets on the basis of the mutual benefit of all member countries, such as the one mooted between Iran, Pakistan, India, and Bangladesh, should be feasible. (12)

The petrochemical industry requires highly advanced technology and large markets, and is, therefore, an area in which cooperation between state petroleum enterprises can be undertaken profitably on a regional, interregional, or global basis. One of the most important products of the petrochemical industry is fertilizers. With the development of mechanized agriculture, the potential demand for fertilizers in developing countries is strong.

Iran has five petrochemical plants but, nevertheless, the country imported nearly half of the 800,000 tons of fertilizers required last year. This reflects some of the bottlenecks which can hinder the establishment of petrochemical plants in developing countries. These include the lack of trianed industrial workers, managers, and technicians and the need for an adequate infrastructure. But the availability of abundant raw materials has led the petroleum producing countries to show great interest in the establishment of petrochemical industries.

It is estimated that 19 Middle East and North African countries will be spending about $15 billion per year in the period up to 1980 on petrochemicals projects. (13) The majority of these petrochemical complexes are being built by OAPEC state petroleum enterprises. As can be seen from the figures published in the OPEC "Annual Review and Record 1976," the OPEC member countries have, either under construction or at the working plan stage, a large number of refineries, petrochemical plants, and LNG plants.

The oil minister of the United Arab Emirates recently issued a warning against the dangers of duplication of large-scale industrial projects, and called for the adoption of a coordinated regional policy. He also pointed out that Saudi Arabia, Kuwait, and the United Arab Emirates have already shown interest in cooperating in this field. As he noted, there is an urgent need for policy coordination between state petroleum enterprises regarding the establishment of new industrial plants. This is especially true for the OPEC state petroleum enterprises, otherwise they may face competition and marketing problems similar to those encountered by the Kuwait Petrochemical Industries Company, which was forced to shut down its ammonium sulphate and sulphuric acid plants.

One of the problems faced by state petroleum enterprises operating in other developing countries is the danger of nationalization in these countries. Most developing countries have nationalized the transnational oil companies on the grounds that these companies had been exploiting their natural resources for many years with the help and backing of the colonial powers. However, the OPEC member countries are making their investments in the developing countries with a view to helping them, and it would be fair if the developing countries were to offer some guarantees to these enterprises, even if they have partnerships with oil companies of the industrialized countries. It is felt that a solution of this problem is possible through reference to international bodies and the Commission on Transnational Corporations of the United Nations.

NOTES

(1) John M. Blair, The Control of Oil (New York: Pantheon Books, 1976), p. 48.

(2) Foad Rohani, "The Iranian Oil Industry, Twenty Years After Nationalization," Farsi Ketabhaye Jibi (Tehran 1977), p. 179.

(3) The oldest state petroleum enterprise in the industrialized countries is considered to be AGIP which was established in 1928.

(4) Raymond Vernon, "The Influence of the U.S. Government Upon Multinational Enterprises: The Case of Oil," The New Petroleum Order (Laval University Quebec, 1976), p. 59.

(5) Mohamed Ali Movahed, Our Oil and Legal Problems (Kharezmi, Farsi, Tehran 1974), p. 126.

(6) OPEC, Annual Statistical Bulletin 1976 (Vienna, September 1977), pp. 62-87.

(7) The Future of the World Economy: A United Nations Study (New York: Oxford University Press, 1977).

(8) Ali M. Jaidah, Opening Address, OPEC Seminar on the Present and Future Role of the National Oil Companies, Vienna, October 10-12, 1977.

(9) "Recent Energy Trends and Future Prospects," paper presented to the U.N. Committee on Natural Resources at its fifth session (E/C.7/70. April 11, 1977).

(10) W.D. Kruger, "Development and Application of Advanced Producing Technology," OPEC Seminar on the Present and Future Role of the National Oil Companies, Vienna, October 10-12, 1977.

(11) Jean-Claude Balaceneau and Jean Favre, "Future Development of Petroleum Research: The Prospect of Transfer of Technology and Establishing Advance Research Centers in OPEC Member Countries," OPEC Seminar on the Present and Future Role of the National Oil Companies, Vienna, October 10-12, 1977.

(12) Financial Times, December 13, 1977.

(13) R. Burchell, "Trends in Processing and the Petrochemical Industry: The Future Role of the National Oil Companies," OPEC Seminar on the Present and Future Role of National Oil Companies, Vienna, October 10-12, 1977, p. 18.

8 Cooperation Among State Petroleum Enterprises in Developed and Developing Countries

Ludwig Bauer

For more than two decades Austria has operated OMVAG, a successful state enterprise which covers the entire range of the oil and natural gas industry, from exploration and production to refining and distribution. After the Austrian State Treaty of May 15, 1955, all property which had been lost after World War II was returned to the Republic of Austria. At the time of its creation, OMV was charged both with developing crude oil production by applying the latest technology, and building up a natural gas industry. The national petroleum enterprise was given jurisdiction over the oil fields situated in eastern Austria and all refineries operating in the country at that time, with the exception of two refineries belonging to the Anglo-American group.

In 1955, Austrian crude oil production more than met domestic demand. Today, Austria can produce only about 20 percent of its crude oil demand, and so it must import oil. OMV has constructed pipelines for transportation of crude oil to the refinery and for moving products to consumers. OMV's natural gas production made Austria the first Western European country to produce as much as 17 percent of domestic gas consumption from domestic wells. OMV was also faced with the task of constructing a modern, efficient refinery to replace the many small, inadequate refineries which had to be closed down. The resulting Schwechat refinery is the largest refinery in central Europe, with a yearly throughput capacity of 14 million tons. Developments have proved that the system existing in Austria provides the conditions for a secured oil supply and, therefore, a secured energy supply position for the country.

As Austria's demand for petroleum products has increased through the years, a dependable import connection with a European port was seen by OMV as vital to the country's needs. This was accomplished through an Austrian branch pipeline connecting the Schwechat refinery with the international Transalpine Oil Line (TAL), running form Trieste to the Federal Republic of Germany. The Austrian leg has a throughput capacity of 11 million tons per year.

To provide for the construction of this branch, contracts were concluded with the transnational oil companies active in Austria: Shell, Mobil Oil, Esso, Total, and AGIP. These six companies own 49 percent of the company operating the pipeline, and OMV owns 51 percent. Processing contracts also permit the transnationals to have their crude oil processed in the Schwechat refinery and enable them to buy additional products to satisfy the market demand.

This fact is of great importance to Austria and could also be of interest to other countries. The arrangements were made in this fashion because competition is welcomed in Austria. It is seen as a challenge always to be better than our competitors. The presence of large corporations forces OMV to apply the most up-to-date processing methods and the latest technical knowledge. OMV and the transnationals apply the same technology and economic efficiency. However, differences do exist in other areas. Ways in which these differences can be decreased or eliminated will be discussed later.

The OMV is not active all over the world, but it is charged with meeting the demand in Austria. Such a charge places a far greater responsibility on state enterprises than that experienced by transnational companies. The transnationals put a much higher priority on generating profits than on assuring supplies. In some cases, the transnationals have simply withdrawn from countries where sufficient profits could not be generated. Italy is one example of this. However, energy supply is the basis for national economic development, and state petroleum enterprises are required to provide the right quantity of oil at the right price at any given moment.

Since the creation of Austria's state petroleum enterprise, the logistics of supply have changed fundamentally. Austria has confronted the transnationals with a strong state petroleum enterprise and has forced them to accept this enterprise as a competitive standard. The continuing competition between the two groups has been beneficial to the energy supply of the country. In the future, however, this situation will change. It will be the cooperation between state petroleum enterprises in consumer and producer countries that will be the determining factor as to whether these two interdependent groups will continue to exist.

In examining to what extent the Austrian experience can be a guideline for other countries, one has to consider the countries that will become oil exporters within the next few years. Included within this group are those countries that are meeting domestic need or are producing somewhat less. At the same time, there are many countries that will never have crude oil production.

As for the countries that will become producers or exporters, in the experience of OMV, the sine qua non of success has been the training and technical capability of the Austrian working force. To achieve similar success, other oil-producing countries must give their workers top training and the opportunity to exchange experiences with those from other countries. The knowledge gained in one country should be at the disposal of those in other countries, and OMV is ready and willing to set an example in this matter.

It is clear that natural gas will be of increasing importance during the coming decades. As growth in crude oil production slows, natural gas will have to be used as a substitute. Doubtless, during the next decades, crude oil will have to be restricted to use in transportation and petrochemistry, and natural gas will have to take over where it can. OMV believes that the state petroleum enterprises will, therefore, have to organize natural gas industries in their respective countries. The OMV and its associates have had great success in the gas sector and have developed the capability to produce, transport, and store natural gas. Bilateral agreements have assured Austria of natural gas imports. The natural gas transportation network constructed by OMV is evidence of Austria's willingness to cooperate in the development of a European gas system. The coming decade will require construction of not only more gas pipelines, but also port facilities for handling liquefied gas. This is a task that demands sophisticated technical expertise; expertise which could be shared with other state petroleum enterprises as it has been shared in the oil sector.

A subject of considerable importance in the framework of a state petroleum enterprise is crude oil processing. As mentioned previously, Austria's one refinery has a yearly throughput capacity of 14 million tons. On the basis of long-term processing agreements, transnationals can have their crude oil processed in this refinery. The fact that there is only one refinery in Austria has made it easy for OMV to operate it alone. This does not necessarily imply that the state petroleum enterprise should own all refineries in a country. In fact, it should be stressed that although ownership of refineries by state petroleum enterprises certainly offers advantages, it does not justify the exclusion of transnationals from refinery ownership. But it is true that nationals should have the ability to exercise technical control, to examine construction, and after completion to operate refineries with the same efficiency as transnationals. Here again, the foremost task appears to be the training of personnel, with all its attendant problems, problems that can be solved only through willingness to cooperate closely.

Naturally, the risk is small for transnational companies that have their processing done for a fee in a state-owned refinery and, therefore, are not required to make major investments. This simplifies the development of financial agreements that are mutually advantageous to the parties concerned. In this connection, it is important to note that state enterprises in the energy sector do have tasks beyond sheer profit-making. It would be totally wrong, however, not to try to run such state petroleum enterprises on a profitmaking basis. The people of these countries cannot be expected to subsidize their state petroleum enterprises out of tax funds. On the contrary, with in the framework of meeting market demand and securing a steady flow of energy, it must be possible to generate profits which could, in the end, aid the country.

It is an old rule that the oil industry should be fully integrated. The ideal goal is to encompass all sectors from production to distribution; and while all countries cannot produce oil, without doubt, this problem can be overcome by cooperation with oil-producing countries.

As to distribution over the long term, the goal must be to supply only the transportation sector and the petrochemical industry. Therefore, a national brand name organization and a network of filling stations should be created in each country. It will not hurt such organizations, however, if they compete with larger transnational organizations. This, again, will be an important incentive for efficiency and, therefore, will be beneficial to the consumer. The experience with such an arrangement in Austria has been good and has created a situation which provides security and independence.

The state petroleum enterprise has a special responsibility to use all available domestic energy sources to achieve the highest possible rate of self-sufficiency. The transnational corporations like to claim that they are in a better position to maintain a steady supply than state petroleum enterprises and, as this statement can only be refuted in practice, the enterprises have a responsibility to produce better results than the transnationals in technology and economic efficiency. These circumstances make it seem natural that state petroleum enterprises should cooperate as closely as possible. Close cooperation could ultimately result in development of a global supply strategy by the state petroleum enterprises similar to the strategy the transnationals have constructed.

This brings us to the question of the areas in which state petroleum enterprises should cooperate. Exploration is one such area. The methods of oil and gas exploration are not controlled by a monopoly and, therefore, are available to all state petroleum enterprises. However, the conditions for development are very different from country to country. Some state petroleum enterprises exist in highly industrialized countries, others in countries with much of their development still ahead of them. If the less developed countries made use of Austria's technical know-how, it could help to bring about an equal standard. In the transportation sector the same approaches should be used.

As far as construction and operation of refineries is concerned, this is an expecially good opportunity for the exchange of experiences to exist. In fact, this has already been proved in practice. Some state petroleum enterprises own only one refinery and may not have at their disposal the comparative data necessary to determine whether or not it is operating at maximum efficiency. If, however, the various state petroleum enterprises compare results, demonstrate opportunities for improvement, and exchange development techniques, the efficiency of refining operations of state petroleum enterprises will undoubtedly reach that of the transnationals.

In the distribution sector, the consumer has the right to expect and get the same advice and service which the transnationals offer. This is why transnationals should be permitted - in a controlled fashion - to be active along with state petroleum enterprises. They will provide a criterion against which performance can be measured and so will force the application of the highest standards of technology and economic efficiency.

The final question is how this exchange of experience and training should take place. It is, perhaps, no coincidence that this symposium is meeting in Vienna, the home of a highly developed state petroleum enterprise that functions in competition with the transnationals. The OMV is ready and willing to put its know-how at the disposal of all who want to make use of it. Let us try to discuss point by point all these proposals, to examine them, and to do jointly whatever is possible. Each country is trying to reach the same goal: to improve its standard of living. No doubt one way to achieve this goal is the development of oil companies that are working for the maximum benefit of their countries.

9 Latin American State Petroleum Enterprises and Their Association in ARPEL

C. Vanrell Pastor

The activities leading to the establishment of ARPEL were not conceived as a solution to immediate economic problems, but as an outgrowth of an idea implicit in the establishment of the Latin American petroleum enterprises: the importance of hydrocarbons as a strategic element of economic life. At the same time, emphasis has been placed on the close relationship between energy consumption and the production of goods and services, which is the basis of progress and of technological development in those countries.

For the sake of clarity, it is desirable to examine the ways in which Latin American governments have intervened in the petroleum industry. First, they have acted in accord with their basic responsibility to guarantee the supply of energy products, both as holders of the relevant mining rights and by virtue of their authority to establish and regulate the exercise of those rights. Second, they have taken steps in accord with their responsibility to perform necessary functions in the public interest, directly or through contracts with third parties.

In fulfillment of this second function, a number of Latin American governments have decided to intervene in petroleum operations either directly or in association with private enterprises. This has been done generally through the establishment of state petroleum enterprises which either hold a monopoly on or take an active part in exploration, production, refining, and marketing.

An association of the Latin American state petroleum enterprises was discussed for the first time in 1961 in a meeting a Maracay, Venezuela. At that meeting Dr. Perez Alfonso, then Minister of Mines and Hydrocarbons of Venezuela, rightly emphasized that the primary objective of a state enterprise is the very important one of promoting the general welfare of the nation. These enterprises, despite their past failings, have created important centers of development and constitute a major driving force in the technological field. The state petroleum enterprises are extremely powerful in their countries. Some have attained international dimensions and have extended their activity beyond their borders to include foreign operations. Generally speaking, they have an adequate financial base from which to promote their own growth. Moreover, the funds generated by their operations make a significant contribution to the financing of public expenditures.

Despite some divergence between the original motives to create these enterprises and the institutional functions and forms adopted, there are common factors which have inspired and continue to inspire this policy. The original motives included a concept of national sovereignty extending to the management and use of natural resources such as petroleum. These resources are considered to be fundamental, both for their strategic importance in the economic life of a country and for their value in international trade. Another factor was the principle of optimization of natural resources, from which each state is bound to draw maximum benefit for its own people.

The state petroleum enterprises, subscribing to the development policy prevailing in Latin America where integration is regarded as the best solution for regional problems, founded ARPEL (Asistencia Reciproca Petrolera Estatal Latinoamericana); and, in so doing, added certain individual features to this general principle. While the established goal of the regional integration machinery was to increase trade, because of the experience in mutual assistance gained before its foundation, ARPEL chose two parallel lines of action: one intended to increase trade in those goods directly or indirectly within its sphere of competence, and the other aimed at raising the technological level of the member enterprises. The basic ideas which inspired direct intervention by the Latin American governments in petroleum operations have evolved over the years. On the one hand, the enterprises have modernized and clarified the concepts which led to their establishment, and on the other hand, they have modified the general trend and specific characteristics of their relations with the international petroleum industry, which has likewise experienced major changes in its dominant role in relation to world energy policy.

The most publicized change has been from an outlook based primarily on principles of national sovereignty to one based on an appraisal of petroleum as a nonrenewable natural resource, in other words to a conservationist view. The question of nonrenewable natural resources is very important for the Latin American countries which have been unable to diversify their resources, or have not known how to do so. The exhaustion of a resource, therefore, influences their decisions and must be taken as an economic factor, as must be the elasticity which exists on a global scale between the supply of a resource and market price levels. The nature of risks involved in mining, production, and recovery of hydrocarbons, the dynamism of the technology employed, and the vertical nature or concentration of operations have convinced these countries that without direct intervention it is impossible to acquire the experience and knowledge required to make the right decisions. The growth of the Latin American petroleum enterprises reflects the typical problems and difficulties of technical development. At an arduous and trying time in history, they were faced for long periods with wide-ranging and varied competition that was experienced and well prepared. This convinced those in many of the enterprises of the need to create their own scientific and technical support as well as selection criteria to systematize the experience acquired and to establish protective measures. Within this technical, commercial, and political framework, ARPEL has helped to crystallize cooperative ideas and aspirations. Its objective, formulated with due regard to the legal and institutional framework of its members and of the association itself, was defined as mutual assistance for the purpose of considering and recommending to the members agreements which would promote their common interests, with a view to the economic and technical integration of Latin America. Its motto sums up this objective: "Towards the integration of the Latin American Government oil companies."

The aforementioned ideas are presently taking shape, in line with the Statutes of the Association (Article 2) "The purpose of ARPEL is to study and recommend to its members mutual co-operation agreements in support of their common interests, with a view to the economic and technical integration of Latin America."

This Article goes on to formulate a series of specific objectives which constitute the guidelines for ARPEL's general policy:

1. To promote the exchange of information and technical assistance among its members.

2. To carry out research to promote:

    (a) Co-operation with international bodies;

    (b) Expansion of the petroleum industry in Latin America;

    (c) Expansion of Latin American industries producing equipment and supplies for the petroleum industry;

    (d) Conservation of hydrocarbon resources;

    (e) Co-ordination of development programmes among its members;

    (f) Development of commercial transactions among its members.

3. The promotion and organization of congresses, conferences and technical and scientific meetings to consider and discuss matters of common interest.

From the outset, an effort has been made to ensure that all resolutions adopted by the assembly are formulated in an atmosphere of freedom and in a practical spirit, with a view to achieving immediate and pragmatic solutions. Indeed, ARPEL recommendations and studies are only effective when they promote common interests, and their mandatory character depends solely on the will of each enterprise, since it is the enterprises which must determine the validity and importance of their interests. This does not prevent their agreements from leading to specific and positive solutions, and this flexibility enables them to take more rapid action as regards integration, bypassing the legal and institutional machinery of the international Latin American integration bodies.

The main institutional components governing the functions of ARPEL as laid down by the statutes, are the member enterprises, the executive body or General Assembly, and the Secretariat. The coordination of the enterprises with the Secretariat has taken shape through the designation of a high-level official in each enterprise to act as liaison officer. This led, in 1975, to the organization of regular meetings of these coordinators. The other organs through which the association carries out its activities are meetings of experts, specialized commissions (established on a trial basis in 1973 and institutionalized in 1976), and executive meetings.

Although the statutes define the responsibilities and functions of the main organs - the Assembly and Secretariat - they naturally cannot provide a specific definition of the institutional policy pursued. Indeed, no such statement has been issued by ARPEL; and its policy, therefore, must be traced through the ideas behind the various actions taken by it. These ideas are apparent in the content of various Assembly agreements, and in papers bearing on organizational matters.

If these are grouped according to the various lines of activity, a distinction can be drawn between the methods and the machinery used. A number of comprehensive studies on the integration of the hydrocarbons market, and the successive conclusions and recommendations of meetings of experts and specialized commissions show clearly that as of 1968 or 1969 the general view was that the work should be carried out on the basis of information and data supplied by experts on the staffs of the state enterprises that could provide a basis of reference for the activities of the enterprises. As for the role of the Secretariat, the main concerns were provisions of technical resources, personnel, and budgeting.

When the ARPEL Secretariat was set up, there were three possible organization schemes: to establish a Secretariat with specialized departments, following the standard models; to entrust the preparation of specific programs to experts specially recruited for that purpose or chosen from among the staff of the enterprises, and to set up, within the enterprises, committees or commissions responsible for the work.

As early as the first regular Assembly, sentiment was in favor of having specialists in the enterprises develop the Association's program of activities and do the work connected with it. It was made clear that the intent was not to cut costs and reduce the budget, but to take into account the desire of the enterprises to participate directly in the work.

This approach has been utilized at various Assemblies dealing with the different aspects of ARPEL's activities, and the fourth meeting of coordinators adopted the same solution.

In considering the actual operation of the Association, it should be noted that the group which initiated ARPEL's operations was an expert meeting, whose conclusions and recommendations were taken as a basis for the agreements of the Assembly. These experts act as representatives of their enterprises within their purview and with their support. As a result of appraisals of the operation of ARPEL by the Secretariat and by the delegates of the enterprises, the most recent trend is to prepare for the work of the Assembly by convening a meeting of coordinators in advance of its sessions. This is a technical committee which considers the implementation of the agreements and the findings of the meetings of experts with a view to coordinating the work of the Secretariat with that of the member enterprises. The institutionaliza-tion of the meeting of coordinators reflects a greater sense of responsibility on the part of the enterprises for the operation of the Association. It may thus be considered that two institutional tendencies or policies, implicit or explicit, have emerged from the statutes and the discussions and agreements of the Assembly.

The enterprises intervene directly in, and assume responsibility for, the Association's programs of work, and the executive nature of the Secretariat is emphasized, but with special stress on the coordination for human resources which are, in most cases, provided by the enterprises.

The activities of ARPEL have followed two courses, one aimed at increasing trade and the other at raising the level of technical expertise. The methods used to achieve the current aims include the increasing systematic exchange of information among the member enterprises, mutual technical assistance, research activities, training courses for the staff of member enterprises, links with international bodies, and informational and technical publications, as well as expert meetings and the regular annual Assemblies of ARPEL.

The annual ARPEL expert meetings are a particularly effective method of technology transfer. The participants are specialists from all member enterprises. Observers from the international bodies with which ARPEL exchanges information are invited to participate when the subject under consideration is of interest to them. It is the Assembly which determines the subject of the expert meetings in accord with the requirements of the enterprises, and the conclusions reached by the experts at these meetings are considered by the Assembly and are used as the basis for agreements subsequently implementd and coordinated by the Secretariat in collaboration with the member enterprises.

Twenty-nine of these meetings of experts have been held thus far, with a total participation of approximately 900 specialists. More than 450 papers, all originals, have been presented.

The results of these meetigs have been extremely encouraging. Not only have they demonstrated an acknowledged high level of scientific and technical expertise, but the exchange of practical experience has been carried out in a frank and open spirit, producing concrete results in technology transfer. These have ranged from the application and transfer of techniques evolved in enterprises and universities in developed countries, to the development of local technology and training of suitable indigenous personnel, as well as introducing local manufacturers to the requirements of the petroleum industry. In this latter area, the state petroleum enterprises have collaborated effectively with the countries concerned.   This has taken the form of establishment of heavy capital goods and services industries, particularly in Argentina, Brazil, and Mexico.   There operations supply both the petroleum and petrochemicals industries with a considerable proportion of their requirements.

The collaboration that has taken place in the area of training is an integral part of the regional effort. Broad assistance in overcoming the problems associated with training of technical personnel have been offered through ARPEL. One of the most outstanding examples in this field is the assistance provided to the State Petrolem Company of Ecuador (CEPE) in training and development of technicians, which it required urgently following its recent incorporation into the industry.

The countries of Latin America are becoming increasingly aware that it is important for them to gain access to technology by a method different from the traditional importation of technical know-how. The policy of the state enterprises in this respect has taken two principal forms, the first is promotion or implementation of standards laid down by the states to regulate the import of technology. This has involved a careful study of alternatives to ensure that the choice made is in harmony with the country's real needs.   The other has focused on establishment of technical research enters.   Generally speaking, these centers have originated in the personnel training centers and technical laboratories and workshops established within the individual enterprises to support their industrial activities.

As a result of such action, there are at present a number of Latin American centers seriously engaged in technical research required for the development of the petroleum industry. They include: Yacimentos Petroliferos Fiscales (YPF) Department of Research and Development (Argentina), established in 1942; the Mexican Oil Institute, founded in 1964; the Investigation Centre (CENPES) established by Petrobras in 1957 and reorganized in 1974-1975 as a specialized research body; and the YPFB Oil Technology Centre, established in 1967.   The most recently established center is the Venezuelan Oil Technology Institute (INTEVEP), which was founded in 1973 and restructured in January 1977 in accordance with the new policy of the Venezuelan petroleum industry.

These centers have followed various courses and obtained widely differing results. Some of them are already producing new technology that is being used in their countries in place of imported technology. Some of this new technology is being successfully exported on the international technology market.

These efforts, and other still at the development stage in the state enterprises in the other ARPEL countries, reflect the keen awareness of the importance of technological development, and constitute essential groundwork for the development of an indigenous technology. Mutual collaboration has been initiated in this field, and although there have been no spectacular results due to the very nature of the problem, experience indicates that once personal contacts have been established, the coordination work is simplified and becomes continuous.   This activity, together with the elemination of prejudices and the decision to take the risks involved in any transfer of technology is bringing the Latin American countries closer to an autonomous development of technology.

The activities of ARPEL have developed by focusing on the industry's basic problems and have, thus, tended to meet the needs of all its members. To that end, specialized commissions have been set up for each of the industry's main concerns. These commissions, for whose organization some of the enterprises are responsible, analyse and develop the ideas and experience presented at the experts meetings and solve the problems posed by member enterprises.

At the present there are five specialized commissions. They are as follows:

Specialized Commission on Equipment, Materials and Services (CEEMASE)

Specialized Commission on Personnel Training (CEFAP)

Specialized Legal Commission (CEJ)

Specialized Commission on Prospecting and Exploitation (CEPROEX)

Specialized Commission on Refining

It should be pointed out that ARPEL has concentrated on the technical side of Latin American integration, although it has also undertaken other activities, particularly in the commercial field, with a view to the achievement of complete integration. For example, many contracts and agreements relating to the hydrocarbons market or refining have been concluded among member enterprises within the framework of ARPEL. However, ARPEL has not confined itself to this aspect of the matter, but has extended its activities in accordance with the objectives laid down in its statutes, and has sought to promote enterprises in the region producing capital goods and services.

It was in this connection that Assembly Agreement IX .1.1 was passed, leading to the establishment of the Specialized Commission to study the industry producing equipment and materials for the petroleum industry in the countries of member enterprises. The commission was to explore whether the manufacturers of such equipment and materials could establish an information center with ARPEL to examine the potential for production and development of new product lines. The short-term objective of this undertaking is to obtain and provide information, and the long-term objective is to encourage manufacturers in the region to seek opportunities to supply the petroleum industry both directly and indirectly through engineering and service firms in the region.

A decision was also taken to publish a catalog of capital goods and services, and the necessary support was requested from the member enterprises at the Eleventh Assembly. The first edition has already been published, listing equipment and materials made in Latin America, together with the engineering and service companies established in the region.

ARPEL has produced positive results, and progress has been made with regard to organizational matters and working machinery. Recently, in an attempt to determine how the member enterprises see ARPEL, the Association established a process for the dynamic analysis of both the objectives chosen by ARPEL and the performance of the machinery set up to achieve those objectives.

The evaluation has shown that ARPEL has created an atmosphere in which exchange of information and experience has become especially flexible, thus facilitating full use of the substantial technical knowledge of the member enterprises. Special reference was made to the reciprocal training of personnel, which has been generously offered. The evaluation also demonstrated a perception that the bilateral or multilateral agreements concluded among member enterprises have been finalized or facilitated both as a result of the information exchanged in ARPEL and because of the personal relations established between executives and technical experts. Both in the Secretariat and in the analysis carried out by the enterprises, there is recognition of the need for a more detailed accounting and objective appraisal of the advantages which ARPEL can offer its members, and of the need for greater publicity concerning the Association's activities at all levels of the member enterprises, particularly in official circles.

It must be acknowledged that, in the hydrocarbon market, trade increases slowly. This was expected by all familiar with the subject who know that developments in this market are determined by factors external to the decision making powers of the enterprises. However, there are lateral aspects of this issue, such as the interchange of capital goods and services originating in the region and technology market. These aspects not only can give rise to considerable trade flows, but also can provide support for industrialization at a highly technical level. But the activities and significance of ARPEL depend on the importance accorded to them by its member enterprises. These bodies are currently preoccupied with a number of problems of such urgency that all other activities must give way to them.

The objectives originally set for the state enterprises were adapted to the situation of each country. In the case of importing countries, the objectives were to reduce fuel imports, minimize the foreign exchange required for such imports, and diversify the source of supply; in the case of exporting countries, the objectives were to guarantee production and maintain reserves; and in both cases, to provide petroleum products suitable for the domestic market at reasonable prices. The policies established by governments and the functions and organizationl methods of the state enterprises have had to be adapted with a view to the attainment of these objectives.

Owing to the lack of experience and the historical circumstances surrounding their establishment, the institutional forms originally adopted by the Latin American enterprises did not, generally speaking, take account of the particular characteristics of the industry; this is the initial cause of their present problems. It should, however, be pointed out that many of these have already undergone favorable changes with time, and others are undergoing similar changes.

Secondly, government intervention in petroleum operations should be based on appropriate policies. In the region, policies have not been fully in harmony with original goals, and occasionally have flaws, such as discontinuity and contradiction in the application of specific measurers, and a failure to adapt energy policies to the changing situation as regards petroleum, although this is not peculiar to the developing countries nor to Latin American countries. Particular mention should be made of policies relating to domestic prices.

Of course, each nation, in exercise of its sovereignty, applies the policies it considers best adapted to the attainment of its established economic and social goals, but an anomalous situation arises if, for example, the rules of governing state activities are incompatible with state economic policy, or if the state assigns a nonremunerative task or policy of a social or national character to a state or private economic activity without explicitly recognizing that fact. In a free or mixed economy such activities are adequately recompensed; otherwise, the activity appears to be showing a deficit and, therefore, seems to be a burden for the country and becomes progressively more inefficient in practice.

Measures or policies of this type, as well as organizational defects and the physical characteristics of their petroleum production capacity, affect most of the Latin American state enterprises and create problems of the highest priority for their officials. These urgent considerations often have the effect of relegating ARPEL matters to the level of priority assigned to them by each enterprise. However, ARPEL itself must provide an answer for these problems as well. Moreover, the short-term problems should not delay work on solving the medium and long-term problems linked to ARPEL's activities.

The effectiveness of ARPEL is directly linked to the flexibility with which it adjusts to changing situations in the industry. This is why far reaching improvements in its systems and machinery have been proposed, in order to accelerate its activities and to give greater continuity to its joint efforts.

These ideas reflect a deep belief not only in the value of this Association, but also in the results of the meetings of technical experts, which have made possible the most effective exchange and transfer of experience and of technological and commercial developments. The proper implementation of these ideas will depend on the vision of the managers of the member enterprises and the persistence and determination of those dedicated to the goals of ARPEL. General Mosconi, the founder of YPF, a Forerunner of our Association, said in 1925:

> One of the easiest things in life is to have ideas and plans, even very good ones; for this a sharp mind and a little imagination suffice; but the most accursedly difficult thing is to take even the smallest idea or plan, organize it, translate it into action and get it moving.

In view of the current petroleum situation, ARPEL needs to adapt bolder solutions. In some cases, this means breaking with traditional models, often inherited or acquired, and running certain calculated risks.   It also necessitates creativity, in order to modify existing methods or find new ones which, being adapted to the circumstances, correspond most closely to the interests of the enterprises.

Thus, as regards trade, ARPEL has suggested that members should participate more decisively in the aforementioned studies of the Latin American hydrocarbons market, since the improvement of the market structure will increase the potential for trade. Particularly favored is a careful and balanced study of the possibilities for manufacturing or acquiring capital goods and services in the region capable of supplying most of the needs of its petroleum industry. With regard to technology, conditions should be created to steer demands in the state enterprises towards a Latin American technology market, in which supply conditions have, in some respects, been created that would permit practical action to be undertaken once certain natural objections of various origins have been overcome. One of ARPEL's fundamental concerns is information systems.   If the information which the enterprises have acquired from their own experience is properly arranged and organized, numerous guidelines will emerge to encourage collaboration in the important area of research and development.

On the basis of ten years' experience with mutual assistance, it is clear that if the joint capacity, knowledge, and potential of member states can be mobilized, a powerful force will have been set in motion that will help to transform the destiny of Latin America in a very short space of time.

10 Cooperation Among State Petroleum Enterprises in Southeast Asian Countries

Piet Harvono

The formation of a regional organization in the southeast Asian region was based primarily on the desire to contribute towards peace, stability, and prosperity in the region. By the most commonly accepted geographical definition, the southeast Asian region is the area comprising ten states: Burma, Thailand, the three Indochinese states, the Philippines, Malaysia, Singapore, Brunei, and Indonesia. Differences in historical and sociocultural background, religion, and economic and political outlook have been minimized in an effort to seek common characteristics to accelerate national development in each state through regional cooperation. However, it has been only since the 1960s that the southeast Asian states, working independently of any outside countries, have begun to think seriously of establishing a regional organization they could call their own.

It is against this background that the Association of Southeast Asian Nations (ASEAN) was founded in Bangkok in 1967 by Indonesia, Malaysia, the Philippines, Singapore, and Thailand - countries representing 75 percent of the region's population. The main objective of ASEAN is to create a peaceful, prosperous, and resilient community through joint efforts, with a view to strengthening economic and social stability in the member countries.

The basic approach to the development of regionalism is through consultation and coordination, leading to integration of thoughts and the growth of a regional consciousness. This approach is of central importance because a regional organization is judged by the degree to which it is able to catalyze common national hopes into a regional aspiration. Although all developing countries have different approaches to setting development priorities, mainly because of heterogeneity in development, technology, and economic infrastructure, common appeal to regionalism may still be found.

Each nation gives priority to rapid economic development while attempting to minimize internal disturbance and external interference. At the same time, mutual concern for physical security requires not

92

only national resilience, internal cohesion, and stability, but also a strengthened regional equilibrium.

In the case of Indonesia, the elements cited by the government as motivating regionalism are the need to pursue a role compatible with the region's geographical location and historical experience, and the desire to ensure that national resilience in each state remains the foundation for any cooperative enterprise. The Bangkok Declaration described the objectives and extent of cooperation among member states, establishing that the states themselves retain primary responsibility for the stability and development of individual activities. The Declaration also stressed that the main areas of collaboration would be socioeconomic and cultural cooperation including joint endeavors, promotion of regional peace and stability, collaboration, and mutual assistance on matters of common interest, including utilization of agriculture and industry, promotion of southeast Asian studies, and maintenance of close and beneficial cooperation with existing international and regional organizations with similar aims and purposes.

During the early years of its existence, ASEAN was primarily concerned with reaching a consensus through consultation, planning, and adaptation. However, the period after the first ASEAN summit meeting in 1976 has been more achievement-oriented. An important step in the work of ASEAN was the recognition of the strategic importance of oil and its role in the development of the economies of the southeast Asian states. Indonesia and Malaysia are the only oil producing members in the region, while the Philippines and Thailand are the consuming members, and Singapore enjoys a unique position as the crude processing center in the region. The crude oil imports bills for Thailand and the Philippines for 1977 are $700 million and $1.1 billion respectively. These figures are almost indentical to the net trade deficit for both countries. While Malaysia has maintained a net trade surplus, with oil accounting for nine percent of its total exports in 1977, oil revenue in Indonesia accounted for 69 percent of its foreign exchange earnings. This has helped increase Indonesia's foreign exchange reserves from as low as $586 million in 1975 to $2,466 million in mid-1977. Singapore, with the third largest refinery facilities in the world after Houston and Rotterdam, yields petroleum products accountable for 26 percent of its total export ($1,479 million) while domestic consumption costs amount to $981 million.

National management of oil-related matters is handled variously by the ASEAN members. In Indonesia, the national oil company, Pertamina, founded in 1957, has the sole authority for the rights of exploitation of national resources. It is also vested with the rights to enter into contracts for exploration and exploitation of oil with foreign enterprises.

In Malaysia, the national oil company is Petronas, established by the Petroleum Development Act 1974 with the exclusive right and power over exploration and development of petroleum onshore or offshore. At the time of its creation, all companies that had been issued exploration licenses or had entered into petroleum agreements were requested to submit all information and records to Petronas for reconsideration.

To encourage greater foreign participation in developing its petroleum resources in 1972, the Philippines amended the 1949 Petroleum
Act and created the Philippines National Oil Company (PNOC) in 1973
as its sole natural oil company charged with undertaking and transacting
the corporate business relative primarily to oil operations and other
energy resources exploitation.   In 1976 the government created the
Energy Development Board to integrate in one body all policy
formulation and regulatory functions relating to energy exploration and
development.   The Chairman of PNOC automatically became a member
of the Board.   This board was superseded in October 1977 by the newly
created Department of Energy.   The Philippines thus became the only
ASEAN country where the policy for development of the different
energy resources is formulated by one centralized government body.

Singapore has no exploration and exploitation activities, but the
government has a share in the Singapore Petroleum Company through
the Development Bank of Singapore. It started its activities with the
construction of a 65,000 barrels per day (bpd) refinery in 1973, to be
upgraded to 200,000 bpd at a later stage.

Thailand did not have a petroleum law until the 1971 Petroleum and
Petroleum Income Tax Acts were enacted. Under the 1971 Petroleum
Act the Thai Petroleum Committee was established, consisting of
various senior members of the government representing the various
ministries and appointed by the Council of Ministers. This committee
has the duty to render consultation, advice, and opinions to the Minister
of Industry in matters such as the awarding and renewing of concessions, payments of royalty and taxes, and the regulation of import and
export of crude oil and its products.

A preliminary meeting among top officials of national oil companies
and observers of government oil agencies from ASEAN countries was
held in Manila in 1975, to determine ways member countries could
contribute towards the progress and prosperity of the region. One of
the results of this meeting was the founding of the ASEAN Council on
Petroleum (ASCOPE).   The ASCOPE Declaration which created the
Council was signed by the heads of the state petroleum enterprise
companies or national oil agencies in the region.

ASCOPE does not come under the umbrella of the ASEAN
Secretariat-General, but it conducts its program and activities independently within the ASEAN concept and informs the ASEAN Secretariat-General of its program and activities.

The aims and purposes of ASCOPE are in line with the objectives of
ASEAN and include promotion of active collaboration and mutual
assistance in the development of petroleum resources in the region
through joint endeavors; collaboration in efficient utilization of
petroleum; provision of mutual assistance in personnel training and the
use of facilities and services in all phases of the petroleum industry;
coordination of information exchange for promotion of methodologies or
for formulating policies within the industry; holding of conferences and
seminars; and maintenance of close cooperation with existing international and regional organizations with similar aims and purposes.

The Council, the highest body of ASCOPE, is comprised of the heads of member country state petroleum enterprises who meet annually to approve past performances and recommend future courses of action. The venue of the meeting is rotated among the member capitals.

An ASCOPE Secretariat was established at this time and, since Jakarta is the seat of the Secretariat-General of ASEAN, it was decided that it should also be the seat for the ASCOPE Secretariat. At this early stage, the Secretariat is primarily a documentary and an information office with no authority for regional level coordination for implementing the approved work program. It is intended, however, that the Secretariat will be charged with the maintenance of a proposed central data bank once it becomes operational.

In each member country a national committee is set up under the chairmanship of a person appointed by the respective council member to implement the decisions made by the annual council meetings and to support the continuing operations of ASCOPE. At present, the national committees meet together twice annually, with the venues of the meeting rotated among the members. The host of the national committees meeting chairs the meeting.

The ASCOPE Secretariat submits its reports of activities to the meetings of the national committees. The national committee of each ASCOPE member country submits copies of the minutes of each meeting to its respective national secretariat of ASEAN for information and coordination of related affairs. This is to ensure that there is no duplication of effort in the petroleum field, either by ASCOPE or by other private or government agencies.

To assist the meetings of the national committees, three working committees comprised of experts and officials were established on specific subjects. These include a technical, an economic, and a legal committee. These committees submit their reports to the semiannual national committees meeting for comments or approval and are directly responsible for the implementation of the approved work programs.

The chairmen of the working committees are distributed and appointed in alphabetic rotation for the duration of one year or more, depending on the assignments given to them. This is done to guarantee equal and full participation of all ASCOPE members in approved work programs.

The technical committee of ASCOPE is charged with identifying, assessing, and recommending areas and projects for regional technical cooperation, assistance to member countries, and other activities that would benefit member countries and regional development. Included in these areas of cooperation are manpower development, exchange of information, petroleum exploration, development and utilization, marketing, holding of symposia, and assessment of facilities available to other members.

The terms of reference of ASCOPE's economic committee include initiating measures to facilitate concerted action among ASCOPE members to raise the level of economic activity in the petroleum sector and to maintain or even strengthen economic cooperation in the petroleum industry; as well as undertaking investigations and studies of the economic, financial, and ecological aspects of petroleum industry development, including joint projects and activities. It is stipulated that these activities should not overlap with ASEAN economic activities.

The functions of the legal committee include preparation of a report comparing the various petroleum laws of the ASCOPE member countries; conducting feasibility studies on the adoption of uniform or standardized policies and regulations to ensure efficient utilization of petroleum and its products; and provision of legal advice when needed.

The oil industry concerns itself with the exploitation of nonrenewable and finite resources often found in isolated areas, the development of which requires heavy investment in infrastructure and facilities. The difficulty of geological and geophysical interpretation cause the industry to operate at levels of risk - financial and political - which are significantly higher than those for other industries. The oil industry is frequently characterized by vertical integration from exploration to marketing. These characteristics become the determining factors in the relationship among the host country (be it developed or developing), those who extract the mineral resource (state petroleum enterprises or the transnationals), and the ultimate users (usually those in the industrialized countries).

The constitutions of Indonesia and The Philippines stipulate that natural resources, including oil and gas, belong to the state. Estimates of the reserves of oil in the ASEAN region vary, but the French Petroleum Institute, basing its conclusions on reports from 18 oil companies, nine independent experts, and two public authorities, estimated that there are 16.6 billion tons of conventional petroleum reserves in the ASEAN region. Taking into consideration the present rate of production and increasing consumption, unless new discoveries are made, these reserves will only last for the next 15 years.

In 1975, oil and gas represented 95 percent of the total regional energy usage. With the development of other energy sources, this ratio is expected to drop to 80 percent in 1985, assuming that the risk capital for the required exploration is available (table 10.1). The higher costs of exploration in later years may indicate the degree of exploration saturation, as easily accessible reserves become more difficult to find. As a rule of thumb the capital investment required per barrel of reserve is $2.00.

Notwithstanding the enormity of the task, and the heterogeneity of ASEAN member countries, efforts at finding a common ground among the ASCOPE members have been fruitful. The approaches taken by ASCOPE to transform ASEAN objectives into realities include the establishment of a lateral and hierarchical structure within ASCOPE; identification of mutual interests and common problems with avoidance of wasteful competition; and an examination of the methods of operation and channels of authority relating to ASCOPE in order to improve the utilization of capital, technological and managerial expertise, and facilities available in the petroleum sector.

The machinery of cooperation among ASCOPE members has included review of existing facilities to identify the potential for common use

TABLE 10.1. Demand for Oil in the ASEAN Region
                   (in barrels per day)

	1975 (Actual)	1985 (Estimated)
Indonesia	248,000	489,000
Malaysia	101,000	186,000
Philippines	201,000	357,000
Singapore	78,000	156,000
Thailand	184,000	350,000
	812,000	1,538,000

and to prevent duplication of construction efforts or purchases. Cooperation has also extended to the review and exchange of existing practices and experiences to identify the potential for standardization of methods, regulations, or specifications; exchange of trainees; and cooperation in selected economic and technical matters. Other activities of ASCOPE have included exchange of views and data of common interest, joint studies on topics of mutual benefit, such as industrial complementation studies, and ad-hoc studies as requested by member governments. ASCOPE has also investigated the utilization of ASEAN machinery for dialogues with industrialized countries or state associations and mutual aid in times of emergency.

Most state petroleum enterprises, including those outside the ASEAN region, have their origin in downstream activities and market outlets. It is, therefore, natural that these fields have been the first targets of regional cooperation. However, in view of the limited availability of state funds for capital - intensive oil development projects, together with the acute shortage of technical know-how and management expertise, those involved in ASCOPE accepted from its very beginning that joint investment projects were not appropriate topics for the early ASCOPE meetings. ASCOPE intends, rather, to approach the mechanisms of regional cooperation through problem solving and conducting studies on topics of interest to all members.

Perhaps the most significant area in which the ASEAN members have undertaken cooperation is that concerning the seas. Offshore exploratory drilling, although introduced in the Brunei/Sarawak waters in 1956 and 1957, came to Indonesia only in 1968. Indonesia, recognizing the importance of the ocean to ASEAN activities, took the initiative early to demarcate its offshore boundaries. In most cases, the boundary demarcations have been settled by agreement among the states, usually using the equidistance rule, although in some cases physiographic features have also been taken into account.

Besides being rich in natural resources, the ASEAN waters are strategic as regards movement of tankers and other commercial vessels. Its vastness makes it difficult to police adequately against pollution and complicates much needed marine research in the seas of the region. Although an inter-ASEAN governmental body has been set up to regulate shipping in the Malacca Straits and to increase safety in navigation, ASCOPE still concerns itself with the problems of oil pollution in offshore areas. Action now under way includes the compilation of governmental oil pollution regulations in effect, identification of government agencies responsible for monitoring environmental hazards, and the identification of research and other activities now being undertaken in each member country in offshore waters that have direct bearing on oil pollution.

As increasing numbers of state petroleum enterprises purchase tankers, the establishment of a standardized operating procedure for the abatement of marine pollution among ASEAN members is a worthwhile effort. Aspects of this problem being considered include offshore safety regulations and response to oil-spills, especially with regard to the selection of dispersants.

To enhance the geological evaluation of the ASEAN region, a joint geological study has been set up. Selective studies have examined the stratigraphic correlation in the South China Sea and the Gulf of Thailand and a geothermal gradient map of ASEAN offshore waters has been prepared.   Other studies now being carried out include stratigraphic correlation studies in the Sulu/Celebes and the Andaman Sea, and preparation of an ASEAN heatflow map.

The technology of offshore exploration, although still related to that employed for land-based operations, is so new and expensive that most offshore activities have been conducted by transnationals only. This, in turn, has necessitated the creation of government regulations covering the range of offshore activities in view of the absence or the limited character of relevant control systems now in place.   To prevent duplication of effort in this area, ASCOPE established contact with the Committee for Co-ordination of Joint Prospecting for Mineral Resources in Asian Offshore Areas (COOP), a regional body established by the United Nations Economic and Social Commission for Asia and the Pacific (ESCAP) in 1966, to which all five ASEAN countries belong. With COOP acting as the offshore data bank for both general and detailed surveys, it is hoped that joint projects can be undertaken to supplement past surveys, to the benefit of the offshore oil industry.

In the field of transfer of technology, ASEAN stresses cooperation in the use of training, education, and research and development facilities through identification of the facilities and personnel available in each member country. This effort puts other members in a better position to identify their training and educational needs.   Exchange of training personnel and trainees has now become a regular feature among ASCOPE members, and in-house courses, largely in the field of oil production, are now attended by other members of ASCOPE.

To further facilitate inter-ASEAN trade, comparative studies have been concluded on fuel and lubricant specifications in ASEAN countries, with the result that there is considerable support for ASEAN specification on fuels and lubricants.   Such a specification would define the agreed testing methods and the accepted ranges of analyses for fuels and lubricants used in the region.   Consideration is also being given to the problem of lube oil recycling in the region, as it is widely thought that governments should exercise some control over the quality of recycled oil marketed by small private companies.

In the economic field, cooperative efforts are being undertaken in product marketing and supply. The long-term petroleum product supply and demand picture for the region is being studied.   In the matter of crude supply to refineries, a study has been initiated to consider the effects of the various ASEAN-country crudes on member-country refineries.   This study is an outcome of the adoption of the Agenda for Emergency Oil-Sharing by the ASEAN economic ministers.

In the legal field, an attempt is now being made to collect all the laws and regulations of ASEAN member countries relating to oil, including environmental protection and petroleum conservation.   It is intended that all the data collected will be stored at the Secretariat, which is charged with the establishment of a data bank.   The legal aspects of this exchange of data on oil among members are now being considered and finalized.

From the beginning ot its activities, ASEAN has utilized the unanimity rule in its decision making process, a policy that has been adopted by ASCOPE. The birth of ASCOPE coincided with the aftermath of the 1973 energy crisis, which has changed some of the basic strategy thinking in the region. For consumers, oil became a very expensive commodity, while for producers, oil is now a major source of foreign exchange to sustain national development. All agree, however, that exploration for oil and the development of known reserves should be intensified, especially in view of the coming of the next energy crisis predicted by many independent reports and studies for the 1990s.

The regional potential for oil has been proved in the past by discovery of oil onshore as in the case of Indonesia. Lately, however, offshore discoveries in the ASEAN region have also been making headlines. With Indonesia having an offshore production above 500,000 bpd and Malaysia around 185,000 bpd, it is certain that greater possibilities for oil are present.

Among ASCOPE members, much progress has been made in identifying common problems and interests in the region. But more efforts will be required before finally succeeding in rallying the divergent national interests to joint action on an agreed program to solve mutual problems. Regional cooperation is a slow, long-term process that requires gradual and periodic adjustments as well as constant efforts from all member countries concerned with sustaining it. Cooperation among state petroleum enterprises can accelerate not only the development of the oil industry in each member country, but also the development of the region as a whole. This development will be instrumental in the build-up of regional resilience, so vital for the survival of southeast Asia as a region, and of each member country individually.

During the first ASCOPE conference a senior official from the Chase Manhattan Bank Group stated that the capital requirements for the decade 1976-1985 for the Far East (including ASEAN) to expand regional development in the offshore areas would be $116 billion. The massive technological and capital requirement of this offshore expansion can be met only through the participation of transnationals and privately owned oil companies. Therefore, extraregional participation will remain a dominating factor for the years to come, and it is up to the representatives of state petroleum enterprises and the government to decide how, where, and when to stimulate greater participation by oil companies in the search for and production of oil.

11 The Necessity for Setting Up a Regional Organization for Mutual Cooperation and Assistance in Oil Matters in Africa*

Recommendations made by the African Delegates

The majority of delegates from African countries attending the United Nations Interregional Symposium on State Petroleum Enterprises in Developing Countries, consulted one another and agreed to remind their respective governments to expedite action in setting up a regional organization for mutual cooperation and assistance in oil matters among African countries as was recommended by the first African Conference on Hydrocarbons, re-recommended by the second African Meeting on Energy, and endorsed by the fourth Conference of African Ministers. The following reasons for undertaking this important step were given:

1. Oil is a very important driving engine for economic progress, be it in industry or agriculture, which is badly needed in Africa. African countries, therefore, should explore all available and most expeditious avenues for effective development of this very essential means for rapid economic development in their region, which can transform the living conditions of the people of this region from developing to developed as it has done in some other regions of the world.

2. Exploration for development of oil resources is a very costly venture, and success of this needs cooperation among those involved in this industry. One of the reasons for the success of the oil majors or big transnational oil companies is the fact that they have always cooperated among themselves in the form of either partnerships or associations in operating in the different sectors of the industry. They share in heavy costs, risks, experiences, information and know-how. Interstate cooperation in the oil industry is a must for success.

3. Regional mutual assistance organizations in oil matters exist in other parts of the world, such as OAPEC in Arab countries, ARPEL in Latin American countries, South East Asian countries,

---

* Recommendations made by the African delegates to the U.N. Interregional Symposium on State Petroleum Enterprises in Developing Countries held in Vienna, March 7-16, 1978

and even in the Western countries through their transnationals. This cooperation has been very useful in developing oil industries in these regions. In the Middle Eastern, Latin American and South East Asian areas that have regional organizations, many countries have already formed their own state petroleum enterprises or national oil companies. In Africa, some state petroleum enterprises are already formed and are functioning, some are formed and struggling to find their way to function, some are in formation, and some are yet to be formed but, nevertheless, are being considered. It is, therefore, necessary to start an interstate body to help these governments and state petroleum enterprises.

4. The United Nations encourages the establishment of such mutual assistance bodies in economic matters. International aid institutions are favorably disposed to giving aid to such regional bodies that cooperate and help member countries and serve whole regions rather than individual countries.

5. Oil fields and oil bearing formations transcend international boundries, so it is economically cheaper to carry out certain oil operations, such as seismic and preliminary exploration surveys, on a regional basis rather than one by one in individual countries in that zone.

6. There are certain problems in oil exploration, exploitation, transportation, and marketing which are peculiar to the African region, and which need joint and cooperative action by all those member countries taking part in the oil industry in the region.

7. Regional cooperation of this sort in economic matters is in keeping with the objectives and goals of the OAU, the ECA, and the UN.

The delegates pointed out some potentially fruitful, viable, and possible areas of cooperation for a regional body.

Exploration

There are many small countries in the region where there are possibilities of the existence of oil - at times even in big commercial quantities - but because of lack of capital and know-how, those countries cannot carry out the preliminary seismic and/or exploratory surveys, the results of which can be used to easily attract loans available from international loan institutions to develop these deposits. A regional cooperative body of the type suggested can pool the individual financial resources of the members to hire or contract companies with the know-how to carry out exploratory or seismic surveys in groups of member countries on a zonal basis at very much reduced costs. Such a body can afford to hire or establish firms with modern and sophisticated exploration technology to be used by countries starting in oil industries, or anyone else needing such services in the region.

It has to be remembered that the major international oil companies are in the business for commercial purposes to make profits in the shortest time and through the easiest possible means. But they are not primarily concerned with national interests. A country may have enough oil deposits in the territory for national needs, so that it may not need to import oil and can, hence, save invaluable foreign exchange. A transnational corporation may not be interested in developing and exploiting such deposits, since it does not serve the profit motive. This, therefore, is the responsibility of that particular nation, helped by the regional organization, to develop the industry.

A transnational is likely to concentrate on exploring selectively very prospective areas which are easily accessible, leaving the other, more difficult, and not rapidly profitable areas. With national interest taken into consideration, both areas would be given considerations for exploration. A regional organization conscious of national needs and interests, and not only of a quick profit, can help a country drill greater depths, both onshore and offshore and in difficult terrains.

Technical Training and Personnel Development

One of the major problems in the oil industry in developing countries is a lack of necessary personnel to operate the industry. While individual countries can build their own institutions for training oil engineers, technologists, and technicians, it is more economical and useful to pool resources and establish larger institutions of such nature on a regional basis. Such regional institutions with greater financial resources can then recruit the best available experts in the field from all over the world to teach young men and women from different parts of Africa in oil technology. Such an institution can attract aid more easily from bodies like the UN, EEC, and other regional petroleum organizations. Such a training institution will have a multiplying effect because the young men and women trained there will go back to their own countries not only to work but to train others locally. Such an institution will enable intermixing of different nationalities and interchange of different views, which will foster African unity.

Joint Ventures in Refining, Petrochemical and Fertilizer Industries

There is no need to elaborate on the advantages and profitability of joint ventures of such industries as refineries or petrochemical and fertilizer plants. Such industries established on a regional basis and planned to avoid wastes, unnecessary duplications, and inefficiency will ensure a ready supply of vitally needed energy. It will also help meet the needs for ever-increasing petrochemical products and vital and essential supplies of fertilizers in Africa. This will greatly help increase food production in Africa.

Oil Transportation in Tankers

Oil transportation by sea is sometimes very risky. But, if African countries have to master all phases of the oil industry, they must learn to operate and run their own tanker services. This can be set up by the regional organization. With these facilities a country can share and spread risks, costs, and profits resulting in better services.

Trans-Africa Pipelines

Construction of joint venture refineries and petrochemical ventures as suggested above will lead to the construction of pipelines for crude oil and refined petroleum products from one country to the other in the region. One such line can be a trans-Africa pipeline running west-east or north-south as the needs may dictate. This will serve all the countries through which it passes, thereby becoming a great instrument for enhancing economic relations between African countries. There will also be smaller pipelines connecting two or three neighboring countries. All these international pipelines can be built and can operate as joint ventures between member countries in the type of regional mutual cooperative organization being suggested for benefits of all.

Research

Useful research into new techniques for oil exploration and exploitation and for techniques that suit particular environments and regions need a lot of money. Such ventures could best be carried out on such a regional basis. It is more economical and effective that way. The member countries of the regional mutual assistance organization being suggested can pool their limited financial and human resources in this field and build a bigger and far more useful institution than individual member countries can do. Such regional research institutions will attract experts and aid from abroad more easily than those serving only one country. The research institutions should establish relations with similar institutions in other parts of the world for exchange of experience and ideas.

Joint Investment Corporation

The regional organization can also set up a joint petroleum corporation which will finance, either through loans or contributions from member countries, other projects in the field of petroleum and petrochemicals. The corporation can also advise member countries on the best way to invest their capital funds so as to ensure optimum financial growth and returns. The project may also be in nonpetroleum fields, but consideration should go first to petroleum and petrochemical projects.

Other Areas of Cooperation

There are many other mutually fruitful areas of cooperation within such an organization, and the others can be elaborated when the regional body is formed.

Modus Operandi

The aforementioned delegates appeal to their respective governments, the ECA, and OAU to expedite consultation on the government level with the aim of forming a regional body for cooperation and assistance in oil matters.

On the basis of 1975 statistical information concerning world exploration programs, only 3 percent of that effort had been put in the 80 nonproducing developing countries, most of which are in Africa. Their conclusion is, therefore, not that Africa does not have oil, but that there is too little effort being made to explore for it. Africa could be harboring as much oil as the biggest producer today.

III

Technology and Training

12 The State Petroleum Enterprise and the Transfer of Technology

D.H.N. Alleyne

There has developed in recent years a clamor for the "transfer of technology" from the developed to the developing countries. This request reflects two significant international developments.

The first has been the efforts of the developing world to engage in certain "sophisticated" types of economic activity formerly confined to the developed countries. When economic activity in the developing world was largely geared to producing raw materials, there was little cry for this "transfer of technology." As soon as the developing world became politically independent and sought to strengthen economic activity by including the kinds of industries found in the developed world, the demand for the transfer of technology became urgent. The second development was the assertion of the principle of permanent sovereignty over natural resources. In 1962, when developing countries gained international acceptance for their assertion that they had the sovereign right to determine freely the use of their national resources (later restated as the right to own, manage, and utilize their national resources), the demand for transfer of technology assumed a new significance.

These two developments gave rise to a situation which has been described as the assumption of an "adversary relationship between the economic power of the international corporations and the political sovereignty of most countries." (1) On the side of the developed countries, the problem has been put in various forms. C.P. Kindleberger's view was that oil production "is highly capital intensive, using a foreign, standardized technology and specialized skills of labour; what little local labour it uses will have difficulty in absorbing these skills and technology." (2) Similarly, the attitude of oil men reacting to the takeover of petroleum resources by a developing country: "Venezuelans can get the technicians, but not even Moscow will share the technology." (3) On the side of the developing world is a cry bewailing the disadvantages of not possessing the relevant technologies and demanding that they no longer be deprived of an essential ingredient for

109

economic growth and development.   "A typical developing country depends technologically on developed countries in a manner that is quite asymmetric, involving a relation of subordination, and it is the asymmetry that makes the notion of technological dependence a central concern in economic development." (4)

Since developing countries asserted the right to manage their national resources, any step to realize this right leads automatically to a reduction or variation of the role of the transnational corporations that at one time dominated the petroleum industry. Hand in hand with this assertion of ownership rights in regard to mineral resources is an equally important assertion of rights by the transnational companies that they must also have the right to own, control, and benefit from the equally important inputs into the petroleum industry - the technology for exploration, production, and refining of petroleum. This technology is often protected by patents, internatioanlly recognized and safe-guarded for the benefit of the original discoverer.   They are also protected by the company itself against industrial espionage. But, this latter right is not only asserted.   It is also invested with the characteristic that the technology exists only because of certain qualities inherent in those who developed it, and that the developing world does not have the capacity to develop that technology - at least not in the short run.

It is not proposed to argue at any length, at this point, the short-comings of this assertion. R. Navarre has effectively rebutted this kind of assertion of a basic and persisting inferiority.   "In general we underestimate a great deal the ability of the most humble minds to interpret and analyse once they are in possession of the information necessary." (5) But this mistake is not confined to the developed world. People in the developing world often assume it and are inhibited by it. Having convinced themselves of their own inadequacy, they destroy their self confidence and, thus shackled, proceed as though there is an all-pervasive something called technology which they do not have, which resides elsewhere other than with them, and which can only come to them if someone else transfers it to them.

What does the "transfer of technology" mean?   To hear the developing countries pleading for or demanding it, and the developed countries or their companies arguing about the difficulty of transfer-ring, or bargaining for a quid pro quo, the impression is gained that there is a fixed pool of something which can be packaged and transported from one place to another. The governments of developing countries ask the governments of most developed countries to arrange for the transfer of technology. The latter respond in a way that gives the impression that they are both dealing with so many tons of some commodity.   But technology is not in any sense a pool or sum of anything finite in quantity or quality. It is being constantly changed and updated.   Also, it is not one thing or one kind of thing. It may involve equipment and tools that can be transferred, but it is more than this. It may involve patents and processes.   There are internationally recog-nized procedures for registering ownership of and rights to patents in respect to equipment or processes, and which determine the duration of the right. It is possible to buy the right to use a particular thing or process. But technology is not only processes and patents, however important these are in the concept of technology.

Technology is really a combination of factors that includes all tools, equipment, processes, and patents. But it also involves a certain intangible additional input - knowledge of an ability to manipulate and use those tools, equipment, or processes for the attainment of specific ends: the production of certain goods and services. This involves human development, a conditioning of the human mind to comprehend certain tehniques and relationships, how certain things "work," and why or why not. In fact, petroleum technology may be regarded as systematic and formulated knowledge about the industrial art in the field of petroleum. This body of knowledge is always changing, being amended, increased and improved by the research, thoughts, and chance encounters of people within the industry or outside it.

The technology can be learned. Learning it can be facilitated by different forms of assistance, teaching, training, and exposure to it. It is doubtful, however, whether it is possible to "transfer" it.

J.K. Galbraith asserts: "Technology means the systematic application of scientific or other organized knowledge to practical tasks." (6) Melvin de Chazeau and Alfred E. Khan, while concentrating upon questions of industrial structure and organization, have given some very interesting insights into the concept of technology that help to clarify the issues and indicate to developing countries that acquiring a given technology is a task which, while it has its difficulties, is not beyond their abilities. (7) DeChazeau and Khan point to three aspects of the process of technological change that need to be differentiated:

1. the true invention - the radical new idea that departs from accustomed ways of thinking and doing,

2. the process of innovation - translating invention into commercially feasible technology or adopting known ideas or methods in new combinations, in new areas and with new results,

3. simple rationalization or scientific management, the application of better methods and techniques to the production process. (8)

They further observe that all of these categories represent aspects of technological progress. In practice, all, but more importantly the first two, are usually interdependent.

From the point of view of most developing countries however, the emphasis at this stage is not generally upon the first aspect above. They are most concerned about acquiring a commercially feasible technology; that is, adapting known ideas or methods in new combinations, and simple rationalization or scientific management.

The further point is made that:

Every invention, however radical, is a re-arrangement of prior knowledge. But at some point it is useful to disinguish: (a) the

significant innovation, conceived as a simple flash of genius from, in
sequence (b) the invention that is the end product of painstaking,
systematic group research (c) the closely related innovation that
brings an invention to fruition or adapts it to more uses and (d) the
mere application of new techniques accessible to anyone skilled in
the prior art. (9)

What has additional significance for the developing world must
necessarily be the "rearrangement of prior knowledge" and having
persons "skilled in the prior art."

Grasping this point is fundamental to an understanding of the
problem facing the developing countries. It can provide the key to its
solution and can indicate one aspect of the role of the state petroleum
enterprises in the transfer of technology - the process of technological
innovation. This process involves borrowing, adapting, and experiment-
ing with ideas already present in our fund of scientific knowledge.

The course to be followed by the developing countries is, therefore,
the creation of an atmosphere conducive to the development of a fund
of scientific knowledge and of persons skilled in the prior art, who,
through their state petroleum enterprise are given the opportunity to
borrow, adapt, and, above all, to experiment with ideas until these
become part of the economically usable technology.

Given the nature of technology and the way it is developed, the term
"transfer of technology" is based upon a misconception. It may be more
meaningful for the developing countires and their state petroleum
enterprises if, instead of asking for the transfer of technology, they
adjusted their thinking to comprehend the nature of the problem and
sought: 1) to acquire the right to use certain aspects of the relevant
technology; 2) to develop within their own spheres of operation the
atmosphere conducive to a familiarization with the basic scientific and
technical ideas and operations relevant to the industry; and   3) the
training of their own personnel in the prior art and the development of
knowledge of the industry.

The second two developments above are crucial. It is essential in
obtaining the optimum benefit from the right to use aspects of
technology, for, while it is possible to acquire both the patent and the
services of imported personnel to employ the patent, unless local
personnel and the right atmosphere are developed, the benefits to the
developing country will cease with the term of the contract of
acquisition.

Reference was made earlier to the possible development of an
adversary relationship between the economic power of the transnational
corporations and the political sovereignty of the host countries. In fact,
most developed countries and transnational corporations are prepared to
make the technology available - but at a price. There is the feeling
that since developing the technology in many cases involved a long
period of time and considerable expenditure, the technology should not
be freely available. It should be exchanged for a fee, or be used as a
bartering counter for certain rights to natural resources that developing
countries own and control but cannot develop without this key of

technology. Engler refers to the representative of the United States in discussions on the transfer of technology, specifically to his view that "restrictions against corporate investment in natural resources were seen as discouraging the infusion of capital and the transfer of technolgy necessary for growth." (10)

The representative of the Federal Republic of Germany at UNCTAD IV made the following remarks on the transfer of technology: "The industrialized countries, too, are interested in intensifying the transfer of technology to developing countries, since the growth of their economies depends to a large measure on opening up such new markets."

Noting that "technology know-how is mainly in the hands of private companies," he pointed out that the direction of the transfer of technology is determined for the most part through their investment decisions.

It is clear to the developing world that, as far as the developed countries are concerned, the transfer of technology is not unidirectional.   There is a quid pro quo.   In return for the transfer of technology, there must be a fee or an opening of markets or investment possibilities.   The governments of developed countries are prepared to help persuade companies to arrange the transfer of technology, but circumstances must be right and will depend on the existence of favorable conditions in the recipient country, such as "agreements on the encouragement and reciprocal protection of investment...." (11)

One of the obstacles to the transfer or the acquisition of technology is the price demanded by the developed world and its corporations for making the technology available.   The insistent request from the developing world must be seen as implying that, in some sense, the price is too high.   At the United Nations Interregional Conference on the Development of Petrochemicals in Developing Countries (Teheran, 1964) reference was made to "the high fees and royalties charged for know-how" and to the fact that often the fees operated irrespective of the size of the project.   An enquiry was raised of the possibility of varying the system in favor of underdeveloped countires.

The noncash aspect of the quid pro quo raises a more fundamental problem for developing countries.   They see the insistence by transnationals and their home countries on direct participation in their economies as an attempt to continue the old domination under a new guise; as an attempt to deprive them de facto of a sovereignty recognized de jure; as a reversal, in fact, of their efforts to achieve political and economic independence and control over their natural resources.

A number of steps by both developing and developed countries seem to be necessary.   Developing countries have to be prepared to pay something for the acquisition of the right to use processes developed by others. More importantly, they must train and acclimatize their people to the new intellectual environment. They must produce personnel with training, experience, and familiarity with the natural sciences and, in particular, with the petroleum industry.   Only in this way can they

ensure that the little they concede will sooner or later win them increased self reliance in technology. Without trained personnel, they
will not be able to use the technology effectively. The developed
countries must realize that any concessions they make as to levels,
fees, or the size and nature of the noncash quid pro quo can be a
profitable investment in terms of goodwill and access to a market
whose growth they foster.

The steps outlined above could be facilitated by reducing corporate
secrecy, whether pursued as national security, proprietary rights, or
technological readiness, and by a conscious effort to lift the veil and
the mystique that surrounds the petroleum industry, for "knowledge
about oil technology and oil policies is not as esoteric as the corporate
stewardship would have the citizenry believe." (12)   Not so long ago,
Japan (now an industrial giant and a leader in many branches of
technology) was derided as it made its first faltering steps to acquire
more advanced technology.

Analysis is in vain without considering the time constraints on the
acquisition of technology. The developing world is in a hurry. It wishes
to compress into one decade or less the technological advances and
adjustments accomplished by others in half a century or more. With
such objectives, one must commit resources and energies to the means
required. Peoples or nations ambitious for advancement, and in a hurry
to achieve, cannot demand that such advances and achievements be
handed to them on a platter. They must consciously prepare themselves
for the objectives they set for themselves. In this connection it is
instructive to borrow, in part, from the contributions of one representative at UNCTAD IV: "We developing countries have deeply realized
from our own experience that independence and self reliance is the
fundamental policy in developing the national economy." (13)

The following section of this paper treats some steps that can
facilitate the technology, bearing in mind that in technological
advances no country is an island.

A first step for the developing country and its state petroleum
enterprises is to discover which companies or firms own or control
various aspects of the petroleum technology. A comprehensive list of
the relevant areas of the technology could be prepared, against which
could be listed the names and particulars of the institutions that own,
control, or do research in these areas. These institutions include the
major companies, specialist companies, research firms, equipment
manufacturers, private consultants, universities, and technical institutes and schools. The United Nations may be able to play a useful role
in starting the collection of that data on an objective basis and without
favoring any particular institution or country. Such data should be
made available to all member countries of the United Nations. If the
United Nations could be persuaded of the need for the establishment of
an International Energy Institute, that body would be ideal for
completing and updating any studies on research undertaken to provide
the kind of information mentioned above. It would also be able to give
assistance in other energy matters to the benefit of both developed and
developing nations.

It is evident that technology does not "reside" with one company or a group of companies such as the Seven Sisters, the group of transnational petroleum corporations with whom the developing countries and their state petroleum enterprises have normally to bargain or negotiate over national control of petroleum resources. In fact, very few processes of interest to a developing country's state petroleum enterprise are owned or controlled exclusively by a major oil company or independent. Such processes, techniques, or patents are usually owned by specialist companies. In most cases, the majors themselves do not have an in-house capacity to do these tasks and so contract the services of the specialist firms for those areas of technology.   In the field of exploration, for example, the processes and techniques are owned by a few specialist firms that undertake aeromagnetic and seismic surveys, fly and map areas, make soundings, and interpret data. There are some small specialist firms that excel in interpretation of seismic data. Some majors have developed and own some of the technology in this area, especially in the interpretation of survey data, largely because they wish to keep secret the results of surveys and tests, so that they can capitalize on any successes in initial prospecting by taking up acreage adjacent to a successful find.

There is available to state petroleum enterprises, through the specialist companies, considerable scope for buying into this area of technology.   The present trend is toward more activity in offshore drilling. Again, as a general rule, the major companies do not own the technology. They "buy" the technology for particular jobs. Techniques for drilling diagonal wells (where several wells are drilled from one platform offshore in a circle around the platform) were also largely developed by specialist firms, not by petroleum companies. Many of these techniques were not developed in or by the petroleum industry at all.   For example, Drake borrowed an idea common in brine wells. Directional drilling was first used in diamond mining in the Transvaal in 1905. Well logging, seabed production techniques, submarine pipelines and storage for oil and gas, and blow-out preventers are not the result of petroleum company efforts. Above all, the technique for extinguishing or controlling well fires is not owned by the majors. One man and his organization have become internationally famous for this kind of operation - Red Adair.

In refining, some major advances were not made by the major companies. The earliest process for desulfurization of crude oil was developed in 1887 by Herman Frasch, a German chemist. The patent was purchased and successfully applied on a large scale by the Standard Oil Company.   Similarly, the Dubbs process for thermal cracking originated with Jesse A. Dubb, a pioneer refiner in California who patented his still and the process in 1909. When in 1913 he realized that his original claim was too modest, he amended his patent - evidently he had developed a process for continuous thermal cracking instead of the batching operation he had imagined at first.   Also, the move to add tetraethyl lead to motor gasoline to reduce knocking in engines came from outside the petroleum industry, from the automobile manufacturers.

These examples illustrate the fact that no company or country has a monopoly on knowledge, especially in petroleum refining. Developing countries and their state petroleum enterprises have some scope for bargaining as they seek to acquire technology. Further, once they become directly involved in the industry, they will contribute to the development and expansion of this technology.

A word of caution is necessary concerning refining. There are firms that offer a package deal for the construction of a new refinery; a company will design, construct, start up, and assist for a period in the operation of a new refinery on behalf of a company or state petroleum enterprise - a turnkey job. In such cases, state petroleum enterprises must have competent staff to act as understudies.

There is also the situation where a developing country nationalizes, with compensation, a refinery. Some of these old refineries were developed piecemeal by the transnational companies. Some portions were designed in-house, the processes being owned by and protected for the company. The equipment and processes met the particular needs of the integrated company according to the crude input, the product mix, or the regional or international requirements of the company. The output of the refinery may have been for direct sale to the final consumer, or for transfer to another company refinery for blending and further processing. Some refineries have idiosyncracies which must be carefully learned by the state petroleum enterprise operators before continuous capacity operations can be assured. Some refineries may have to be altered to meet the requirements of a new owner. In such cases, it may be advisable for the state petroleum enterprise to negotiate with the transnational for a service agreement, the length of which depends on the level and quantity of personnel available to the state petroleum enterprise.

The technology involved in lubricants is special in a certain sense. Each company, by research and investment, has developed special blends of lube oils to meet various industrial, climatic, and other requirements. The technology and the lubricants are special because the makers of high performance machinery, after exhaustive testing, designate a certain make and brand of lubricant as that which must be used for that particular piece of equipment. In addition, insurance companies insist, as a condition for coverage, that a particular brand of lubricant be used and in the manner specified. State petroleum enterprises will, therefore, find that unless the question of force majeure is introduced, they are committed to buy the process and must negotiate with its owner. This does not mean that they cannot make or blend lubricants for ordinary use. The question is whether, initially at least, they can defy these considerations or whether they can afford the expense of trying to produce various blends of lubes in relatively small batches.

Developing countries and their state petroleum enterprises will find, therefore, that, as they need to employ the technological key to the development of their petroleum resources, they are not subject to the same bargaining pressures as would have been the case had they been negotiating with one company or a group of companies that not only dominated the oil industry but also controlled major aspects of the technology.

With the reversion of national assets to the developing country or its state petroleum enterprise, an offer might be made by the major to negotiate with the specialist company on behalf of the state petroleum enterprise for continuation of certain services, especially those for which a period contract had existed between the major and the specialist company. Each case has to be determined on its merits. State petroleum enterprises may find that they can negotiate for themselves, or that, if they accept the offer of the major to act as an intermediary, they should ensure that a representative of the state petroleum enterprise participates at every stage of such negotiations. This is an essential part of the learning process. It adds to their own body of knowledge and expertise, and contributes to the training of personnel in such fields as petroleum economics, making them familiar with the full range of activities involved, giving them a better feel for the industry and its complexities.

However, not all technology has to be bought. In many areas (for example in production), the normal technology is largely free, and it is no longer covered by patents nor is it attributable to a particular company or person.

When, in this matter of transfer of technology, the state petroleum enterprise pays a fee to a specialist firm, normally it is paying for: a peice of equipment or a tool; a process; a technique; or a service which involves the know-how - the expertise necessary for the effective employment of the tool, process, or technique. It is this latter item, the service know-how or expertise, that poses initially the greatest problem for the developing country and its state petroleum enterprise. Only by training its own personnel can it start making the real breakthrough into technological advances.

The task of ensuring that state petroleum enterprises acquire the technological key to the development of their petroleum resources while benefiting from foreign assistance cannot be left to governments of developed countries or to international organizations alone. The basic requirement is for the developing country, acting through its state petroleum enterprise (itself cooperating with certain indigenous or foreign institutions of learning), to train its nationals and selected personnel of the state petroleum enterprise in acquiring the knowledge necessary for operating the various aspects of the petroleum industry. In this exercise, they will find that the governments of some developed countries will be eager to assist. At UNCTAD IV, the representative of the Federal Republic of Germany made the following observations about direct government-to-government cooperation in the field of technology transfer:

We feel that openings for more direct governmental activities are to be found rather in areas where the scope for cooperation is essentially beyond technical and industrial programs. I am thinking, for example, of:

- assistance in developing independent research capacity in the developing countries;

- support given to national or regional technology centers;

- the utilization of research capacity in the Federal Republic for the study of problems of the technological development of products and techniques suited to the requirements of the developing countries or their adaptation to these requirements. (14)

For their own part, developing countries and their state petroleum enterprises must take the initiative to create the necessary institutions for promoting education and training in the natural sciences and in petroleum technology. They must organize and attend seminars, symposia, and conferences on the petroleum industry. They must arrange attachments for their personnel with companies and institutions to expose them to the operations of the oil industry and to the way in which research is undertaken and pursued. They must utilize portions of the income for research and development geared to petroleum technology. Some state petroleum enterprises are specifically required or encouraged by the provisions of their founding statutes to do this, as the following examples indicate:

- The Nigerian National Petroleum Corporation is specifically empowered to "train managerial and technical staff for the purpose of the running of its operations."

- The National Iranian Oil Company is required to provide education and training for its employees.

- The Trinidad and Tobago National Petroleum Company is empowered to initiate training programs designed to ensure that nationals are provided with the training, qualifications, and experience necessary to equip them for attaining the highest positions in the petroleum and petrochemical industries.

- The Iraq National Oil Company statute provides for:

   1) the establishment of training centers for the workers;

   2) the organization of seminars to acquaint personnel with the latest developments in the oil industry and the management of companies and industrial projects;

   3) the enhancement of work in the company by introducing a system of work incentives ensuring payment of bonuses to personnel who submit studies or inventions which further the company's objectives or who excel in their work performance and in raising output;

   4) the establishment of educational and technical institutes to study all aspects of oil industry, company management, and industry project work;

5) the establishment of laboratories and research centers;

6) the sending of educational and technical missions outside Iraq... for specialization at universities and educational institutes...and the training in projects and factories concerned with production and processing of oil;

7) allowing other organizations in the public and private sectors to benefit from the training centers; and

8) establishment of a vocational and petroleum education organization...to undertake the attainment of these objectives.

● In the case of Ente Nazionale Idrocarburi, 15 percent of its profits must be set aside for research and training.

It would be interesting to ascertain to what extent these statutory provisions have been realized or where similar arrangements have been made without the statutory requirement.

Not all state petroleum enterprises in developing countries will be able to establish training institutions. There is, therefore, considerable scope for cooperation in this area. Some countries are already well prepared to offer this type of assistance to fellow developing countries and their state petroleum enterprises (Table 12.1).

2   The oil industry has not always been sophisticated. The base from which the present industry developed, rapid as the development has been, was simple and pedestrian, as a critical historical review demonstrates.

In training personnel, therefore, the state petroleum enterprise must take a balanced view. It must train top-level scientific and professional personnel both at home and abroad in the best possible institutions of higher learning. However, it must consciously train at all levels to develop what R. Navarre calls the "third element" (after raw materials and industrial equipment) of the prosperity of developing countries, i.e., the largest possible number of technologists trained to attain the highest possible level of efficiency.

W. Arthur Lewis' definition of "technicians," a minimum for these purposes, is: "persons who after receiving a basic education to 15-16 years of age, have received the equivalent of 2 years of full-time technical training - whether in a full-time institution or whether over some longer period in some combination of on-the-job training or part-time study." (15)

It is at this basic level of the technological ladder that the shortage is most acute - hence, the need to emphasize the training of drillers, roustabouts, general handy men, refinery operators, or laboratory technicians. Hence, also, the need to stress parity of esteem and adequate financial rewards for all cadres engaged in the petroleum industry, to attract the youth of the nation into this industry. This is also necessary if, once trained, they are to be retained in their national industry and not seduced away by higher rewards in the petroleum industries of foreign countries.

Table 12.1

COUNTRY	SPE	INSTITUTIONS
Iran	(NIOC)	Petroleum Laboratory at Rey, University of Teheran; University of Shiraz; Abadan Institute of Technology
Mexico	PEMEX	Mexican Petroleum Institute
Saudi Arabia	PETROMIN	College of Petroleum and Mineral Resources
Venezuela	CVP and Petroven	Zulio University, Maracaibo; Qriente University, Puerto La Cruz
Iraq	INOC	University of Baghdad (College of Eng., Dept. of Petroleum and Minerals Eng.); INOC Training Centre
Egypt	EGPC	Cairo University; Suez Institute of Petroleum and Mining Engineering
India		Institute of Petroleum Exploration at Dehra Dun*
Indonesia	Pertamina	Technical Institute at Bandung and University of Indonesia
Bolivia	YPFB	Petroleum Development Centre*
Turkey	TPAO	Petroleum Development Centre*
Nigeria	NNPC	University of Ibadan (Institute of Applied Science and Technology)

*Established with the assistance and support of the United Nations.

The petroleum industry is not only fascinating for the ambitious youth of a developing country. Contrary to the view expressed by Kindleberger, it has been described as "a marvelous school for training labor, not only through the combined qualities of knowledge and dexterity which are developed, but also by imposing the twin disciplines of vigilance and rapidity of action." (16) Further, as the examples of Mexico, Iran, or Venezuela indicate, it is an industry on the exapnsion of which it is possible, with proper planning and execution, to rely for the advancement of the economy and the society as a whole. It can be the basic growth point for the economy and the cadres trained within it can, and should, penetrate both physically and by example into the rest of the economy.

The role of the state petroleum enterprise in the acquisition and development of technology was only partially fulfilled when it succeeded in buying the right to use patents developed elsewhere. Its real contribution started when, operating with the tangible support of its government, it established or fostered the establishment of appropriate training institutions for its personnel, at all levels. The next step came with the realization that the character of its personnel and of those whom the latter would inspire in the rest of the society would only have been developed by testing with risky decisions. After suitable training and exposure, therefore, it is necessary to develop character, confidence, and self reliance by permitting the young to borrow and adapt, to experiment with ideas, to innovate and, thereby, make a useful contribution to the technology of their petroleum industry and of their economy, the characteristics and requirements of which they would come to know more intimately and more feelingly than would foreign experts, however competent.

NOTES

(1) R. Engler, The Brotherhood of Oil (Chicago: University of Chicago Press, 1977) p. 144.

(2) C.P. Kindelberger, Foreign Trade and the National Economy (New Haven: Yale University Press, 1962), p. 203.

(3) Engler, The Brotherhood of Oil, p. 34.

(4) Thee, Morek, eds., "UNCTAD IV and the New International Economic Order," Bulletin of Peace Proposals 7,3 (1976): 222.

(5) R. Navarre, "Basic Scientific Surveys and the Training of Men", Proceedings of the United Nations Interregional Seminar on Techniques of Petroleum Development, New York, Jan. 23, to Feb. 21, 1962 (Port of Spain: UNDEXPRO), p. 344.

(6) J.K. Galbraith, The New Industrial State (Boston: Houghton & Mifflin, 1967), p. 12.

(7) Melvin De Chazeau, and Alfred E. Kahn, Integration and Competition in the Petroleum Industry, Petroleum Monograph Series, Vol. 3 (New Haven: Yale University Press, 1959), p. 281.

(8) Ibid.

(9) Ibid.

(10) Engler The Brotherhood of Oil, p. 144.

(11) Thee, Morek, eds., "UNCTAD IV," p. 238.

(12) Engler, The Brotherhood of Oil, p. 220.

(13) Thee, Morek, ed., "UNCTAD IV" p. 240.

(14) Ibid.

(15) W. Arthur Lewis, Development Planning: The Essentials of Economic Policy (New York: Harper and Row, 1966).

(16) Navarre "Basic Scientific Surveys" p. 344.

13 Research and Training in State Petroleum Enterprises

V.V. Sastri

A nation is strong and prosperous in proportion to its possession and application of scientific and technical knowledge. Oil and natural gas are not only vital fossil fuels necessary to turn the wheels of industry and to move men and material, but they are also essential raw materials for fertilizers and manufacture of petrochemicals and other necessities of modern living. Petroleum is thus a strategic mineral.

At previous meetings on related subjects, M.G. Krishna has dealt ably and extensively with the research and training requirements of the developing countries in the fields of refining, processing, and petrochemicals. (1) Therefore, this paper will limit itself to the exploration and exploitation aspects of the petroleum industry in developing country state enterprises, based on India's experience. The basic aim of research in the petroleum sector is to reduce overdependence on external sources, to create a national scientific base, and to establish an organic link with the industry. A scientific base is essential for the balanced growth of any industry. This is especially true for the oil industry, which demands new scientific ideas for such highly specialized activities as exploration and exploitation of hydrocarbons. Constant appraisal and reappraisal of scientific data constitute the basis for the calculated risks involved in petroleum exploration.

The adage that "oil is found in the minds of men" holds true even today. This situation has created a cadre of specialists, the petroleum explorationists. This group is credited with the discovery of petroleum, though ultimately it is the drill which confirms the presence of oil and gas. The petroleum industry is dynamic, with many features that distinguish it from other major industries. It is internal in character, and it requires dynamic management that is capable of taking calculated financial and technical risks of high magnitude. The industry has deep roots in science and technology, and it requires originality in ideas and creative concepts as well as people with the ability to stimulate scientific and technological ideas and development. Workers in the petroleum industry must be familiar with the use of sophisticated instrumentation and of machinery with a high degree of automation. Further, today's petroleum industry can be characterized by versatility and flexibility in process technology and product patterns as well as a diversity of products with many applications. Finally, the industry must maintain the ability to provide well organized and highly developed technical service to customers.   Such a technologically-oriented industry requires well trained and competent technical manpower together with a strong research and development base. A good system of basic and university education and an industrial infrastructure are essential for providing the technical manpower required for the progress of research and development activities in any country.

Occasionally, alarming reports are heard regarding the future availability of petroleum.   A recent energy study prepared by 35 representatives of governments, the oil industry, and science under the auspices of the Massachusetts Institute of Technology, presented before the workshop on Alternative Energy Strategies, predicts that the world oil shortages could begin as early as 1981, if OPEC nations maintain their current level of production. (2)   This panel urged countries to move with "wartime urgency" in developing other fuels and conserving energy. Because most governments and businesses plan only five to ten years in the future, the "basic danger of the world energy situation is that it could become critical before it seems serious." In the wake of such forecasts and warnings, the necessity to discover new oil and gas fields and to increase the recovery factor from known fields assumes increased importance today. For this purpose, the need to initiate and accelerate research and training activities in most of the developing countries (both oil exporting and oil importing) exists.   The United Nations and its agencies, which have the welfare of the world at heart, have a significant role to play in this area.

A recent survey carried out by the French Petroleum Institute on petroleum exploration in 85 petroleum importing market economy developing countries noted that adequate finances for exploration and trained manpower are the main problems facing these countries. (3) The analysis pointed out that most of these countries have some oil either onshore or offshore. It was also pointed out that almost all those developing countries use oil as their most important source of energy, and that only five (India, Republic of Korea, Turkey, Zambia, and Mozambique) consume more coal than oil.   These 85 countries were categorized into two groups: Group A including those with a good chance of becoming self sufficient or even oil exporting, and Group B including those with no substantial production at present.   Group A includes five countries from Latin America (Argentina, Brazil, Chile, Mexico, and Peru) and five countries from Asia (Afghanistan, Burma, India, Pakistan, and Turkey), while the rest of the 75 countries were included in Group B.   The magnitude of the problem of discovering sufficient hydrocarbons in these five Asian countries, which have 40 percent of the world's population, implies the need for large-scale funding for petroleum exploration.   This, in turn, may imply that sufficient R & D activity has not been generated in any of these 85 countries.   The need for basic and applied research in petroleum exploration and exploitation and for the transfer of technology is apparent from such regional assessments. Before discussing the specifics of these problems, some general observations on the oil industry in developing countries will be useful.

Historically, the oil companies have been based in the industrially advanced countries and have carried out various activities in developing countries by importing technology and skills while nationals were trained to carry out routine jobs. Only that much of the technical base which was absolutely essential was established in the developing countries. Few efforts were made to establish research and training centers in developing countries. Nor was there much interaction in these countries between the national universities and research institutions and the oil industry. The basic and applied research and development work was carried out in the companies' parent countries.

An entirely different situation exists today. Many developing countries, and some developed countries as well, have now set up national state petroleum enterprises to look after petroleum affairs. Some of them are involved in petroleum exploration and exploitation. To cite one instance, even though India has been producing petroleum since 1890, it was only in 1956 that the Oil and Natural Gas Commission (the country's state enterprise for exploration, drilling, and production of hydrocarbons) was established. Today, the Indian Oil and Natural Gas Commission employs 24,000 people; 12,000 in scientific and technical occupations. It operates 72 deep and workover drilling rigs, 16 geological and 36 geophysical field parties. The Oil and Natural Gas Commission explores and produces oil and gas on land and offshore in India and abroad. The Indian Oil Corporation, a public sector organization, is concerned with refining and marketing of petroleum products.

Generally speaking, there are two categories of developing country state petroleum enterprises: those in petroleum-exporting countries and those in petroleum-importing countries. Consequently, there will be a slight difference in their outlook and approach. But basically, as far as research and training are concerned, the problems faced by the two categories of state enterprises are almost the same. State petroleum enterprises in developing countries have some or all of the following functions:

1. to advise their national governments in the development of the national petroleum industry;

2. to explore, produce, refine, and market petroleum and its products internally, either independently or jointly with nonnational oil organizations;

3. to supervise or control the activities of nonnational oil organizations operating within the national boundaries;

4. to explore, produce, refine, and market petroleum and its products independently or jointly with an international oil company or with another country's state enterprise; and

5. to assist other countries in the development of their petroleum resources, as mutually agreed upon by the respective governments.

The scope of research in developing countries will necessarily be related at first to local problems. In the first few years, the research activities may not necessarily lead to striking results, because great advances have already been made by major international oil companies in the developed countries. But the developing countries should not be reluctant to face such a situation. When a state petroleum enterprise starts functioning, a wide spectrum of problems ranging from exploration to market applications starts multiplying rapidly. This increases the magnitude of work as well as the sophistication of research. If developing countries want to export products, quality control and related application research assumes great importance. Even if they want only to sell crude oil, the product has to be evaluated thoroughly in the light of modern refining technology to bring the best price. Petroleum has to be evaluated not merely as a source of energy but also as a source of petrochemicals. The content of research is dependent on the scope of operations of the petroleum industry in the developing countries, which is related, in turn, to the current and future industrial activities of the countries concerned. Once this is established, it is then easy to understand the complexity of the petroleum industries and, consequently, the type of research and training support that should be organized. It is now the policy of many west Asian countries to establish complex refineries and petrochemical plants and export processed products rather than crude oil to the developed countries. This, therefore, implies that their research efforts will have to be considerably augmented or initiated, as the case may be, to meet the exacting demands of the consuming countries. Even if the developing countries initially import modern techniques for exploration, drilling, production, refining, and petrochemical processing, the sophistication of indigenous research and service work for quality control would have to be of a high order.

Research and training programs to be carried out by state petroleum enterprises are strongly linked to, and dependent on, the general scientific and technical educational system and standard in a country. Language is also an important factor. The petroleum science has made rapid progress in English and in Russian, French, and Japanese. In order to catch up with the state of the art, the nationals of the developing countries at the university level have to learn at least one of these languages. The research and training component of the petroleum industry cannot develop in isolation. In some cases, the foreign oil companies operating in west Asia prior to 1950 had to establish technical training centers and support the local educational system and facilities. The magnitude and scope of research and training in state enterprises varies from country to country depending upon both the geological setup and geoscientific complexities of that particular country, in terms of exploration, drilling, and production, and the agricultural and the industrial levels as far as refining, fertilizer, petrochemical, and other product patterns are concerned. The west Asian countries export their products to suit the requirements of the importing countries and so the organization of the refineries is tailored not to suit their own countries but others.

The first step for any developing country is to assess its national petroleum potential, in order to plan to reduce its import bill and, thus, become self-sufficient in crude oil production. The basic issue is that the geology of any country is best studied by the country's nationals. However, as geology knows no political boundaries, scientific cooperation and exchange of information between neighboring countries is necessary for mutual benefits. Where prospects of finding oil are limited, some developing countries have taken a bold strategy of exploring for petroleum in other countries. This strategy has necessitated establishment of R & D and training facilities in their own countries deficient though they may be in petroleum resources.

In any case, a developing country that explores for oil in its own or in other countries must have its own R & D base, established either by its own efforts or with the assistance of international organizations such as the United Nations agencies or by mutual institutional agreements with other developing countries. The areas for research in exploration, drilling, production, refining, and petrochemicals include geology, geophysics, geochemistry, drilling technology, reservoir engineering, production practices, refining, downstream products, organic chemicals, general engineering, storage, transport, instrumentation, environment (including pollution), and safety.

Research and training activities are interdependent, as are problems of exploration, drilling, reservoir, and production. Similarly, problems of refining and the required products are interrelated. If the experience gained by India is any guide, problems of exploration and reservoir and production engineering can be solved by a multidisciplinary approach at one integrated research and training institute. After some time, when the scientific and technical problems increase in magnitude and complexity, separate research and training institutes for exploration, drilling, and reservoir engineering can be established.

The tasks of an integrated research and training institute for geological, geophysical, reservoir, drilling, production, and training activities should be to:

- generate basic geological laboratory data through sedimentological, paleontological, palynological, and other microfossil studies;

- interpret geological maps, aerial photographs, and satellite remote sensing for an understanding of regional geotectonics;

- undertake basic evaluation of different sedimentary areas on a continuing basis;

- evolve improved geophysical techniques and interpretation methods for exploration of stratigraphic traps, pinchouts, and so on, and for seismic prospecting in difficult terrains;

- carryout geophysical laboratory studies on outcrop and core samples, and to obtain basic geophysical parameters for correlation and formation evaluation;

- conduct geochemical and hydrogeological studies both in the field and in the laboratory on the origin, migration, and accumulation of hydrocarbons;

- experiment in the field and laboratory on unconventional methods for detection of hydrocarbons;

- study and work out optimum drilling conditions for different formations that are being drilled;

- carry out studies for improving drilling rates;

- prepare drilling plans for exploratory wells;

- solve drilling mud and cementation problems;

- obtain basic data on reservoir conditions (PVT studies);

- study problems on fracturing, acidization, and so on;

- undertake studies on problems of pressure maintenance and secondary and tertiary recovery of oil and gas;

- recommend locations of key exploratory wells;

- prepare plans for exploration and development of oil and gas fields;

- prepare technical and economic reports for development of oil and gas fields;

- advise on oil and gas reserves;

- advise on the exploration and development programs, based on reinterpretation and reassessment of scientific and technical data;

- computerize geological, geophysical, drilling, reservoir, and production data;

- organize orientation and refresher training courses for the scientific and technical personnel of the enterprise; and

- prepare technical manuals and documentation notes to keep the scientific and technical personnel of the state enterprise apprised of the latest technical advances.

The institute should be a think tank for the state enterprise on scientific and technical matters. It should serve as a data bank for all scientific and technical information, and a repository for all rock and fluid samples. On some matters, the institute can act as a technical secretariat to the management. The institute should not be involved with day-to-day technical and operational problems but with persistent problems that cannot be solved in the field.

In order to accomplish these tasks, an integrated institute should have the following laboratories, sections, groups, and units:

Sedimentological, macro- and micropaleontological and palynological laboratories

X-ray, D.T.A. scanning electron microscope laboratories

Basin studies units - each unit consisting of geophysicists, geologists, and geochemists

Delta studies unit (wherever needed)

Seismic, G.M. interpretation groups

Geophysical laboratories

Geophysical model studies unit - consisting of geophysicists and mathematicians

Oil well logging laboratories

Organic and inorganic geochemical laboratories

Hydrogeology section

Pore pressure studies unit

Marine geoscience unit (wherever needed)

Core studies laboratory

Petroleum chemistry laboratory

Technoeconomic study groups

Reservoir engineering laboratories

Formation fluid laboratory

Reservoir modeling group

Production engineering laboratories

Drilling mud and oil well cement laboratories

Experimental drilling unit

Digital computer unit

Computer software groups

Drafting, photomicrography section

Instrumentation unit for maintenance of sophisticated equipment

Mechanical workshop

Library, documentation, and translation sections

Lecture halls, audo-visual aids, auditorium, hostels for trainees

Many developing countries have been carrying out research and training activities directly or indirectly in some form or other at many centers within their countries or abroad under arrangement (see appendixes 2 and 3). But there is also a need to initiate and carry out organized research and training activities either within the state petroleum enterprise or at other centers in the country. The results of such organized research and training activities will positively be felt by the petroleum industry, as has been the experience of a developing country like India.

In 1962, organized research and training activities were established in India with the help of the United Nations and the United Nations Development Program (UNDP), within the Indian Oil and Natural Gas Commission (ONGC). This was the result of a regional international seminar on petroleum exploration and development organized by the United Nations when the need for the establishment of a research and training institute in petroleum exploration, drilling, and production was recognized. The government of India and the United Nations then jointly established the Research and Training Institute, now known as Institute of Petroleum Exploration (IPE), which started functioning in December 1962 at Dehra Dun, the headquarters of ONGC. This institute has grown to become a leader in this part of the world, and it is fully owned and operated by ONGC. During the last 15 years, research and training in petroleum geology, geophysics, geochemistry, drilling mud and cement, synthesis of mud chemicals, formulation of oil-well cement, studies on crude oil, reservoir and production engineering, and drilling engineering have been carried out under one roof. As of 1979, IPE devotes its attention exclusively to problems in exploration. The work is organized into six groups:

1. Research and development work in geology, geophysics, and geochemistry;

2. Basin studies;

3. Exploration problems;

4. Geophysical instrumentation;

5. Computer services and programming; and

6. Orientation and refresher training in geology, geophysics, chemistry, drilling, reservoir and production.

Scientific problems are tackled at the Institute in a multidisciplinary approach by its 500 scientific and technical personnel.

Since the complexity of reservoir, production, and drilling problems has increased considerably, in 1975 the ONGC decided to set up a separate Institute for Reservoir Studies (IRS), as an in-house research organization at Ahmedabad, the center of a major oil-producing province in India. This institute, which started functioning at the end of 1977, concentrates on basic and applied research in reservoir engineering, secondary and tertiary recovery processes, and development of oil and gas wells. Reservoir and mathematical modeling groups are provided with computer terminal facilities in this institute. A total of 150 scientists and engineers work with the IRS in the following laboratories: polymer laboratory, water flooding laboratory, imbibition and wettability laboratory, tertiary recovery laboratory, thermal laboratory, miscible gas process laboratory, acidizing laboratory, well bore problems, PVT and BHS laboratory, core and well logging laboratory, and the petroleum chemistry laboratory.

An Institute for Drilling Technology (IDT) is also being established by the ONGC at Dehra Dun, as another in-house research and training establishment to specialize in drilling problems and technology. IDT will provide a strong base for successful deep drilling. Work is planned in drilling technology, drilling fluids, cementation and cementing material, drilling complications, accidents, and economics of drilling. This institute will employ test drilling rigs, mathematical modeling groups, a documentation center, and a mechanical workshop. The institute may also monitor certain deep exploratory wells.

The Indian Institute of Petroleum (IIP) was established in 1960 by the government of India in collaboration with the French Petroleum Institute as a national research and training center. It, too, is located in Dehra Dun and it deals with problems of refining processes, downstream products, product evaluation, and marketing. The IIP is not an in-house research unit of the industry, but rather it caters to the needs of Indian refineries and petrochemical plants. Its work is complementary to that of the IPE. The IIP has a complement of about 500 scientists, technologists, and engineers. The work is organized under crude oil refining and processing; petrochemicals; projects, product evaluation, and application; engineering services; and training, coordination, and information services. There is also an in-house research and development center owned by another state enterprise, the Indian Oil Corporation (IOC). This center deals with the marketing of refined petroleum products. It was established in 1972 and is located at New Delhi, the headquarters of IOC. It specializes in product development, mechanical testing and tribology, and analytical studies. More than 100 scientists and engineers are working in this center.

Management of a research institute is different from management of a factory or a project where work goes on in a scheduled fashion and quantified targets and measurable indexes are available. Management of qualified scientists, technologists, engineers, and specialists is an art. Each problem has to be handled from a multidisciplinary approach and not according to administrative divisions. Therefore, it is an art of bringing together persons of different disciplines, aptitudes, and views and forming groups to solve the problem on a specific time schedule. This primarily involves creation of a suitable environment under which research can proceed. A certain amount of freedom of thought and action for scientists and engineers is necessary. Mere administrative control and supervision will not produce the desired results. There is no standard formula for the management technique, and it differs from country to country depending upon the attitudes of the people and their working style. The prime function of the manager of a research and training institute is to infuse scientific temper and enthusiasm among the persons working in the institute, and to create a suitable environment for scientists and technologists to work and grow. The head of the institute should be a specialist in his own line, with a good understanding of the needs of the state enterprise, and someone who is able to form groups of scientists and engineers capable of solving problems.

The progress and status of any science-based industry is directly proportional to the degree of sophistication it has achieved. This is all the more true in the petroleum industry. The various facets of science, technology, and engineering involved in the oil industry require specialists of a high caliber. The research and training institutes provide a home for these specialists to grow and contribute to the industry. The petroleum industry cannot progress in any country without specialization in the fields of science, technology, and engineering.

Personnel policy in a research institute is a very important area that should be given careful thought. Since a research institute is usually part of a large organization, its overall personnel policy has to be compatible with that of the parent organization. However, a certain amount of deviation form the general personnel policy of the organization is desirable in the selection and career planning of personnel for a research institute. Those hired have to have an aptitude for scientific and research work. They should be well qualified and have a good academic record. Care should be taken that the researchers are not divorced from the realities of the situation that exist in the field or refinery where their results are to be applied. At the same time, the scientists and technologists in a research institute have to concentrate on a certain amount of basic research in order to solve practical problems. As such, a suitable blend of good researchers and practical scientists has to be maintained. For this purpose, it is desirable to have a core of scientists, technologists, and engineers permanently located at the research institute, with a specified percentage of the institute's total technical strength set apart for periodically bringing in scientists and engineers from the operational areas to work on specified programs in collaboration with the permanent cadre of scientists and engineers. This will ensure the active involvement of the operational personnel with the scientific work being done in the institute.

Various research programs, basic and applied, to be carried out by the research institute alone or in collaboration with national or international research institutes have to be identified and formulated, preferable in consultation with the technical persons who are directly involved in the field operations. This will ensure that the research results have a practical bias and applicability, and provide for the overall utility of the research project. A further step can be taken in which the field operational units can "own" some research projects, in the sense that they can suggest projects for execution. This tactic would be especially useful in some countries where the managers of operational units are relatively uninterested in what is going on in the parent organization's in-house research establishment. Additionally, the research institute has to take up some projects, both basic and applied, on its own initiative. Basic or fundamental research, depending on the country's resources and requirements, should account for about 25 percent of an institute's total work.

There is no fixed formula for funding of research and training activities in a state petroleum enterprise. This has to be viewed against the level of scientific and technical infrastructure that exists in the country. To cite an instance of a developing country like India, where the scientific infrastructure is reasonably good, the ONGC spends about two or three percent of its revenues on in-house research activities. To generalize the situation, it may be said that the lower the standard of scientific infrastructure in a country, the greater should be the input for in-house research. It can be argued that developing countries should allocate at least three to five percent of their revenues from the sale of oil and gas for research activities.

There is considerable scope for collaborative research between a country's state petroleum enterprises and its research institutions and universities. The industry's research needs can be integrated with the country's overall science and technology plan. India has successfully done this. The National Committee on Science and Technology set up by the government of India worked in collaboration with the ONGC in identifying a number of research projects to aid land-based and offshore petroleum exploration and exploitation in India. This program established collaborative efforts between the Indian research institutes and universities, over a period of 5 to 10 years. Over 200 projects in basic and applied, and long and short range collaborative research efforts were identified, covering exploration, drilling, and production. Out of these, 49 projects have been selected as priority projects, and these are currently in progress. Funding is provided by the ONGC. A certain amount of research work, or in a few cases all of it, has been farmed out to Indian research institutes and universities. The IPE is the coordinating organization for these projects. There is also scope for taking up collaborative projects between two or more countries on an institutional basis. This can be between developing and developed countries.

In considering incentives for research, the home governments of developing countries can grant financial concessions, such as exempting expenditures on research and training activities in calculation of income tax to be paid by state enterprises, or exempting imported equipment for research work from customs duties. Similarly, the international oil companies operating in the developing countries could be made to invest a portion of their expenditure or earnings for research and training activities in the developing country jointly with the state enterprises.

Scientific officers working in research and training establishments may also be given certain incentives to do research of a high standard and impart training of a high caliber. These incentives can range from extra remuneration commensurate with the quality of research and training and recognition of merit and work, to better pay scales or better avenues of promotion, or career prospects, in keeping with the personnel policy of the state enterprise. Sometimes, even a certificate of appreciation of an individual's work may serve the purpose and trigger greater dedication of researchers and teachers to their jobs.

There are a number of areas of exploration research that could be undertaken by the state enterprises of the developing countries to suit their local requirements. These include basic studies on the origin, migration, and accumulation of oil and gas; geochemistry of the ancient and recent lacustrine and fluvial sediments; study of the modern and ancient deltas; geological and geochemical relationship of oil and coal-bearing sediments; geotectonics of grabens and foredeeps; study of recent marine sediments of the continental shelf, slope, rise, and deeper parts of the oceans; marine geosciences; occurrence of oil and gas at depths greater than 5 km; exploration for subtle and stratigraphic traps; unconventional methods of exploration; improvement of techniques for acquisition, processing, and interpretation of geophysical data (seismic, G.M., logging); improvement of drilling rates; drilling technology to drill deep wells beyond 5 km on land; research on drilling mud and oil well cement and cementation; direct methods of detection of oil and gas, both at shallow and deeper levels; increasing the recovery factor for oil; and theoretical studies on global tectonics and petroleum.

The ultimate goal of research is to find more oil. To meet this objective, reserves of oil and gas must be increased by either discovering new oil and gas fields, adding annually to the existing reserves at least as much oil and gas as was taken out of the reservoir during the previous year, or increasing the recovery factor for oil from the present world average of 30 to 35 percent to 50 to 60 percent. In the context of a possible world petroleum shortage in the near future, oil should be conserved for manufacture of chemicals and fertilizers, and alternative sources of energy sought. For this purpose, state petroleum enterprises may diversify their research efforts to locate and develop, for example, geothermal energy sources.

The oil industry is capital intensive. Manpower is an immeasurable and valuable asset, but one that may not appear on the industry's balance sheets. In a capital-intensive industry, the quality and ability of the managerial and technical personnel has to be of a high order. It is, therefore, of the greatest importance that state petroleum enterprises recruit qualified and able people and train them adequately to suit the requirements of the organization. Training is a continuous process. Even the senior managerial and technical staffs of an oil organization should be given periodic in-house refresher training courses, even though there may be a certain amount of reluctance on the part of senior officers to participate. Unfortunately, many managers of oil organizations tend to give a low priority to training. There are a few enlightened oil organization managers who have the dynamism and foresight to give training its due importance.

In general, an integrated system of university education, together with a good industrial base, helps to create a technically trained work force in any country. In some of the petroleum-producing developing countries, to a limited extent, the technical base was developed by the international oil companies and the national governments. Thus, there is a close link with the state of the art of scientific and technical education in any country.

Training programs, in addition to imparting knowledge, should try to infuse motivation among the trainees. However much one is trained, if he is not sufficiently motivated to carry out his duties for the betterment of the organization, he will be a burden instead of an asset.

The first level of technical training is for skilled technicians and operators who are mostly undergraduates who pass through vocational training institutes. The second level is for scientists and engineers who pass through universities. Training for the supervisory staff consists of three aspects: formal education in college, university, or technical institutions; orientation training given by the state enterprise, after recruitment; and refresher training given by the organization to its own scientists, technologists, and engineers.

The following different areas and levels of training for the petroleum industry can be identified. It should be noted that not all the areas mentioned are of the same level or standard.

1. Areas of training
   a) Exploration (geology, geophysics, geochemistry, computer studies)
   b) Drilling technology, techniques, and engineering
   c) Reservoir engineering
   d) Production engineering
   e) Transportation and marketing of crude oil
   f) Refining
   g) Petrochemical and fertilizer processes
   h) Plant operation, maintenance, trouble shooting, safety
   i) Marketing of products
   j) Technical service to consumers
   k) Management of men and material
   l) Petroleum policy and planning

2. Levels of training
   a) Scientists, technologists, and engineers
   b) Engineers and technologists for operation, quality control, and trouble shooting
   c) Middle and senior management
   d) Economists and administrators
   e) Officials for management
   f) Marketing personnel
   g) Operators with process and operational knowledge
   h) Skilled technicians

Scientific and technical vocabulary is not well developed in the local languages of a number of developing countries. This requires that the technicians and skilled workers be trained in foreign languages. A number of developing countries may be facing difficulties in communication. It appears that the college of Petroleum and Minerals in Saudi Arabia adopted English for instruction since English is the major world language of petroleum industries and the language of the bulk of the technical literature. On the contrary, Saudi Arabia has established a special institution for development of technical vocabulary in local languages, with the help of United Nations agencies. This situation, to some extent, comes into play at the level of university education. Thus, while communication and language problems may exist, solutions can be found to overcome them.

Research institutes, either within the state petroleum enterprise or outside it, are the best places to impart training. Sometimes theoretical training can be arranged in collaboration with a country's educational and research institutes, but on-the-job training can be provided only by the industry. Training can also be imparted in foreign countries, either under bilateral institutional arrangements or through international organizations such as the United Nations and its agencies.

A comprehensive list of the training institutes in the developing countries where facilities exist is not readily available, but it is believed that some of the petroleum research and scientific institutes of the developing countries listed in Appendix 2 provide some training facilities. The IIP conducts training courses for Indians and foreign nationals (on bilateral arrangement) in various aspects of refining, petroleum products, petrochemicals, and marketing surveys.

The Indian Oil and Natural Gas Commission's in-house training programs have covered a wide variety of topics. The ONGC established an in-house training center within the Research Institute in 1962. Since then, this training center has been conducting orientation and refresher training courses for scientists, technologists, and engineers in petroleum geology, exploration geophysics, geochemistry, reservoir and production engineering, and drilling technology. At the orientation level, the trainees and fresh recruits holding a Master of Science degree in geology, geophysics, physics, electronics, chemistry, or chemical engineering, or a B.S. degree in mechanical or petroleum engineering. The orientation courses run 9 to 12 months and the refresher courses 6 to 12 weeks.

The training center has also organized specially designed orientation training courses for scientists and engineers of state petroleum enterprises of friendly countries under mutual agreement. Under this scheme, the Nigerian National Oil Corporation, Iraqi National Oil Company, and the governments of The Philippines and Tanzania have participated in these courses.

Foreign assistance in research and training implies cooperation and collaboration among the developing countries, between the developing and the least developed countries, and between the developed and the developing countires. This cooperation can be achieved either by bilateral arrangements or under the aegis of international organizations like the United Nations and its agencies, and implies sharing of resources, experience, and expertise for mutual benefit. Some oil-exporting developing countries have adequate finances, while some oil-importing developing countries have experience and expertise. It is necessary to identify and assess the areas, scope, and content of such cooperation. This requires considerable understanding and trust between the concerned parties. Cooperation between developing countries can be very useful because it helps them understand better each other's needs and psychologies.

It is up to the developing countries to decide what they want and how they want to develop their industry. The United Nations agencies can act as catalysts in identifying the needs and initiating a dialogue between the concerned countries. In some cases, national governments acting jointly with the international agencies such as the UNDP, have established research and training institutes. In the case of the IPE, this has taken the form of an in-house research and training institute within the country. Institutional collaboration is another method that takes place between the IIP and the Institut Francais du Petrole. Foreign assistance has, in many cases, accelerated the establishment and growth of research and training facilities, but in the final analysis the nationals themselves have to continue and carry out these activities to suit their country's requirements. Foreign assistance can only act as a catalyst to initiate and establish a base for research and training activities in a country.

An important area of cooperation between developing countries is the exchange of technical personnel to conduct training programs. Personnel who have acquired more skills in some developing countries can work and train personnel in other developing countries. Another area is joint research and development programs. It is often the case that utilization of locally available raw materials and techniques is not of interest to the advanced countries. Sometimes, certain areas of research and development not considered important by the developed countries can be relevant to the developing countries. This should not be misunderstood as an attempt to utilize second rate science and technology. It is a matter of selecting methods suitable to situations that may exist in some developing countries.

The high cost of oil and gas, combined with diminishing reserves, necessitates intensification of efforts to discover and recover more oil and gas and produce at an optimum level.

However, establishment of research and training facilities requires a considerable financial outlay. Alone, many developing countries cannot afford such investments. International organizations like the UN agencies and petroleum-rich developing countries can invest funds in the needy developing countries or regions. It is not suggested that petroleum research and training facilities should be set up in all countries, but rather that regional or subregional centers would be valuable. The existing research and training facilities in the developing countries should be reviewed and their future needs assessed.

It is evident that developing countries need more science and technology for their development. The state petroleum enterprises may have to invest at least three to five percent of their earnings in their in-house research, development, and training activities.

Research and training programs have to be formulated by each state petroleum enterprise depending on the country's needs, local conditions, and available skills and talents. Training should be integrated with the national educational system. Collaborative research may be encouraged with the national universities and research institutes.

External assistance from developed countries and through the United Nations and its agencies will be useful. Such assistance should conform to the policies of the national governments, and institutional collaborations would be preferable. Efforts may be intensified to bring together the developing countries to share their existing resources, facilities, expertise, and capabilities. They can undertake joint research and development projects and training of technical manpower for mutual benefit, since they understand each other's needs and psychologies better. The United Nations and its agencies may establish centers for transfer of technology at selected locations in different regions of the developing world. The feasibility of establishment of an appropriate forum on a permanent basis to encourage various facets of collaboration between the developing countries may be considered. The UN and its agencies may review the existing research and training facilities in all the developing countries and assess their needs. Such an assessment may be made more comprehensive by regional meetings.

NOTES

(1) M.G. Krishna, "Training, Research and Development Needs in Petroleum Refining in Developing Countries," in Developing Countries," in Proceedings of the United Nations Interregional Seminar on Petroleum Refining in Developing Countries, Vol. I, January 22,- February 3, 1973, New Delhi, pp. 254-71; and "Cooperation in the Training of Technical Manpower in Petroleum in Developing Countries," Petroleum Cooperation Among Developing Countries, (New York, 1977) (United Nations Publication Sales No. E.77.II.A.3), pp. 93-102.

(2) Workshop on Alternative Strategies, Energy: Global Prospects 1985-2000, Massachusetts Institute of Technology, 1977.

(3) J. Favre, "Exploration in Petroleum-Deficit Developing Countries," Petroleum Co-operation Among Developing Countries, pp. 21-42 and Corrigendum.

IV
Investment and Financing

14 Investment and Financing Aspects of State Petroleum Enterprises

Arturo del Castillo Perez

The international petroleum industry developed and became consolidated in the opening decades of the present century. In keeping with the pattern of world economic development, it was the industrialized countries that created and applied petroleum industry technology, directly exploiting resources both in their own territories and in their colonies and developing countries, and it is to those countries that the great transnational corporations that to this day control the international oil market belong.

The accelerated growth in world demand for hydrocarbons brought about the invention of the international combustion engine, and the changeover from coal provided the pioneer companies with a broad and dynamic market. The discovery of large deposits of oil near the surface in the developing countries helped their economic, political, and financial consolidation. In addition, the development of new technologies in the various activities of the petroleum industry was conducive to wider and more efficient use of hydrocarbons, so that they became the major source of energy and the main force behind the development of mankind that has occurred in the twentieth century.

Through their control of petroleum technology and the international market in hydrocarbons, transnational corporations obtain concessions to exploit the deposits in developing countries and constitute a force which exerts a tremendous influence on the trend of world economic and social affairs. These corporations manipulate the prices of hydrocarbons and overexploit petroleum deposits to maximize their short-term economic benefits, with no concern for the kind of rational exploitation that would permit maximum recovery. In short, they are vertically and horizontally integrated, covering all phases of the industry, and even setting up financial institutions to provide abundant, convenient, and cheap financing for their exploration, exploitation, refining, transport, and other activities. This control limits, and sometimes prevents, the establishment of state petroleum enterprises in developing countries, since the corporations hold concessions over the most promising geological tracts.

141

The first state petroleum company to be set up in Latin America was Yacimientos Petroliferos Fiscales (YPF) of Argentina, established in June 1922 to avoid problems in supplying the Argentine armed forces with hydrocarbons. The second enterprise to be set up in the region was Petroleos Mexicanos (PEMEX), constituted on March 18, 1938 with the property and concessions expropriated from the transnational oil corporations which had operated in Mexico since 1901. This step was taken because of the corporations' refusal in 1937 to comply with the national laws then in force.

Overexploitation of the petroleum resources of developing countries, the meager economic and social benefits which those countries have derived from the operations of transnational corporations, the political pressures exerted on the governments of the latter, and the fact that the lowest priority has been given to meeting the needs of the domestic markets of the producing countries have all led to the establishment of state petroleum enterprises. The development of such enterprises has encountered endless problems, mainly as a result of the discriminatory policies established for the transfer of technology, the scarcity and inflexibility of international sources of finance, and the restrictions on the establishment of independent channels for the proper international marketing of hydrocarbons.

The main purposes of setting up state petroleum enterprises in developing countries were to give those countries political stability to achieve rational exploitation of their nonrenewable natural resources and to ensure that priority was accorded to the supply of hydrocarbons needed for their economic development.

In a second phase, the establishment of this type of enterprise in some other countries started with the taking over of the petroleum exploitation concessions that had been granted to transnational corporations. Once OPEC had been formed and began to operate, the oil-exporting countries were in a position to set prices for the hydrocarbons they supply to the international market.

The low price which prevailed for petroleum until 1973 and the technological advantages it offered in comparison with alternative energy sources led to accelerated growth in world demand, since the creation of industrial and transport technology presupposed the use of hydrocarbons. Thus, the developing countries, as importers of technologies from the industrialized countries, depend to a large degree on hydrocarbons to meet their domestic energy requirements.

In the first phase of development of the world petroleum industry, which saw the creation and integration of the large transnational corporations, the methods of exploitation used, and the fact that the corporations, controlled the world market and, therefore, the pricing system, meant that there were no financing problems, since the corporations themselves had the capacity to carry out planned investment projects. The discovery of abundant deposits near the surface made exploitation, transport, and marketing highly profitable. In cases where exploration proved unproductive, the governments of developed countries established tax mechanisms which considerably reduced the risks involved in such operations. In that phase, the only parties encountering financial problems were the small companies which tried to survive side by side with the large corporations. In the "easy petroleum" phase, many of them did achieve substantial profits, but eventually they were taken over by the corporations.

When developing countries establish state petroleum enterprises, they encounter a number of problems in obtaining adequate and convenient financing, owing to the fact that the petroleum industry is highly capital-intensive in terms of unit cost. The ratio of investment/value added is approximately 35 percent, while in most industrial activities it varies between 20 and 25 percent. Moreover, most of the financing offered by the international capital market is made conditional on the purchase of goods and services in the countries granting the credits.

In the last two decades, international financial institutions have granted limited credits to developing countries for various activities related to the promotion and construction of oil installations, such as refineries, petrochemical plants, pipeline systems, oil shipment terminals, and so on. However, these credits were made conditional on international competitive bidding for the planning and design phases, for construction, and for the procurement of the necessary materials and equipment, so that the national input into projects financed in this way is small and the indirect effect is to limit the development of industries auxiliary to the petroleum and petrochemical industries in developing countries.

In view of the high level of risk associated with exploration and exploitation, there are no prospects whatever of help from the international capital markets. This being so, state petroleum enterprises have resorted to various methods of financing those phases of the industry. These have ranged from the channeling of government tax revenues or the use of their own scanty resources to the extreme measure, in view of the constraints which exist, of entering into risk contracts with foreign corporations. Such contracts provide for large concessions of potentially productive geological tracts, in order to increase national hydrocarbon reserves. This system, which had been severely criticized in the past, and which the governments of developing countries had abolished, has been used increasingly in recent years because of the financial problems and technological difficulties involved in finding deposits far below the surface. Risk contracts have been prohibited under Mexican law since 1958, in view of their adverse economic, political, and social effects.

In order to meet their expenditure and investment requirements, state petroleum enterprises use both domestic and external sources of finance. Domestic sources include government tax revenues, self-financing, the domestic banking system, national stock markets, and suppliers of goods and services. External sources include international agencies and private institutions, which include the financial system, stock markets, suppliers of goods and services, insurance companies, and pension funds. Which of these sources is used depends on the varying characteristics of the different phases of the petroleum industry, so that each of them must be discussed separately.

In considering domestic sources, the most obvious is government tax revenues, on which state petroleum enterprises have often relied to finance their operations. This involves subsidized prices for hydrocarbon derivatives on the domestic market, which means disregarding any relations between costs and prices and maintaining "policy prices."

This policy rests on the economic fallacy that lower prices for basic inputs are an essential factor in stimulating economic activity. What actually happens is that consumers adopt practices which waste energy, following the same patterns as are prevalent in some highly industrialized countries. The inefficient use of energy in productive activities is encouraged, since the cost of hydrocarbons represents a very small proportion of total operating costs. On the other hand, no stimulus is given to investment since, even when capital formation is boosted in the business sector, a large proportion of the resources involved is diverted to the import of luxury goods for a tiny minority of the population. Thus, the maintenance of "policy prices" constitutes a transfer of state resources to privileged groups and, at the same time, encourages the inefficient use of hydrocarbons.

Considering the prices obtainable for hydrocarbons in the international market, the state petroleum enterprise should be financially self-sufficient, since there is a logical relationship between costs and sales prices in domestic markets. The multiplier effect of investment in the petroleum industry would provide a real impetus to economic activity, help to increase government tax revenues, and possibly stimulate the creation and development of auxiliary industries, a potentially important sector of manufacturing.

In Mexico, "policy prices" also apply to some hydrocarbon derivatives, especially those like kerosene which are used by the lower-income groups. However, a certain balance between costs and prices has been sought so that the state petroleum enterprise may grow with autonomy and without the use of tax revenues, and, indeed, it may become an important source of finance for the public sector. In the federal government's income budgets for recent years, income from taxes on the basic petroleum and petrochemical industry represent 15 percent of total government revenues. This is the result of a discriminatory tax system, a tax of 16 percent being imposed on the sales price of refined products while basic petrochemicals are taxed at 12 percent of the sales price. It should be noted that, under the Mexican tax system for industrial, commercial, and service enterprises, most of the burden falls on profits.

In considering self-financing, it should be remembered that the standard intention for any economic activity is the attainment of some economic, social, or political benefit to justify investment or expenditure. In the case of state petroleum enterprises in developing countries, what is sought is not economic benefit per se; rather, the aim should be to ensure that productive and commercial activities are such as to permit growth commensurate with the increasing national demand for hydrocarbons and, if possible, the export of products to finance imports of goods and services which are not produced in the country.

It is understandable that state petroleum enterprises should resort to national and international sources of finance in order to satisfy domestic needs for hydrocarbons. However, it is clearly necessary that the credits be channeled into profitable projects, according to an order of priority based on the requirements of the petroleum sector. Otherwise, it will not be possible to meet financial obligations on time, and the state will have to contribute to their amortization with tax revenues.

In the case of Petroleos Mexicanos, the enterprise's own resources are channeled into the financing of projects which are essential, but not very profitable. In addition, in order to meet the financial needs of some activities, such as exploration and exploitation for which it is not feasible to seek help from the capital markets, capital reserves have been set up to ensure that they can be carried on in the future. For example, there is a reserve for exploration and oilfield depletion that was created to provide self-financing for petroleum exploration. The basis of this scheme is the imposition of a levy on each barrel of oil produced, the amount of which has varied as the costs of exploration, drilling, and oilfield development increased. This reserve was established in 1945, when a levy of $0.057 per barrel was set. In 1974 it was increased to $2.45 per barrel, and in 1976, because of the large output of the oilfields in the southeastern part of the country, the levy per barrel was reduced to $0.97. As of December 31, 1976, this reserve amounted to $1.21 billion, from which adequate financing can be made available for these activities.

It should be noted that, because it is financially solvent, Petroleos Mexicanos can gain access to the domestic and international capital markets without needing a government guarantee, since it uses the credits only as a supplement for its own funds. In most developing countries, the domestic banking system has only limited funds available for lending, while the needs of the state petroleum industry, because of the size of the investment required for its projects, preclude financing from that credit source. Accordingly, the domestic banking system can cover only a small percentage of the requirements of state petroleum enterprises, and this will be in the form of short-term and medium-term credits. Although in the case of domestic financial sources the interest rate is higher than it is in international markets, the risks arising from changes in the rate of exchange when international currencies are involved are absent. Petroleos Mexicanos uses financing from the banking system, in the form of short-term revolving credits or medium-term (three-year) credits, mostly for domestic purchases of goods and to cover the local costs of construction projects.

Stock markets in developing countries are generally small; accordingly the issuance of bonds by state enterprises is not a common practice, and it has been done on only a few occasions. At the beginning of 1977, Petroleos Mexicanos issued bonds in Mexico, redeemable in four years. They pay a rate of interest lower than the prevailing market rate, but the basic price of the bonds will rise as the price of oil increases on the international market between now and 1980. The upward trend in world oil prices was a tremendous incentive to Mexican investors, and the issue of "Petrobonds" was entirely taken up within a few days.

State petroleum enterprises also use credits from domestic suppliers of goods and services, but for the most part these are very short terms, since they are normally financed in local markets where high interest rates and short repayment terms are the rule. Hence, these sources of finance meet only a small proportion of the credit requirements of state petroleum enterprises.

The development of the state petroleum industry in developing countries depends upon access to foreign sources of finance. These have undergone considerable changes during the 1960s and 1970s; however, they still do not offer adequate terms for the proper development of these enterprises. Since the 1973 crisis, international financial agencies have been allocating limited resources to the petroleum industry in developing countries. However, the finance provided by such agencies (IBRD, Eximbank and the regional financial agencies) has been channeled mostly into industrial and transport projects promising a high return, and do not contribute to a solution of the capital problems with which the state petroleum industry and the economic development of developing countries are beset. This is partly due to the fact that the system of "subsidized prices" for hydrocarbon derivatives makes a number of projects not sufficiently profitable.

International agencies do not finance petroleum exploration activities, which, because of the high risk involved, have a highly uncertain outcome. The resulting lack of financial resources has compelled state petroleum enterprises to enter into "risk contracts" providing for large concessions of geological tracts with oil potential to transnational petroleum companies, which have enormous financial and technological resources as well as the benefit of tax systems in their countries of origin that enables them to reduce the risks inherent in such activities. Petroleos Mexicanos has not had access to this source of finance.

If state petroleum enterprises are to develop satisfactorily, financial mechanisms will have to be created to enable exploration and exploitation activities to be stepped up in order to increase their reserves and output.

Since the establishment of the Latin American Energy Organization (OLADE) in 1972, one of the aspirations of the member countries has been the formation of a regional financing agency which would help to solve the financial problems confronting the energy industry in Latin America. It has not been possible to bring this project to fruition for the following reasons:

1) There are no properly evaluated energy projects making it possible to determine the financial requirements of the region.

2) The more highly developed countries in the region have seen no advantage in setting up an agency of this kind because their economic structure gives them direct access to international financial markets.

3) All the countries of the region, except Venezuela, have serious financial problems and are not in an economic position to contribute the very sizeable capital necessary in view of the characteristics of the various energy projects.

In these circumstances, it would seem desirable, as a first phase, to establish funds in the various geographical regions where the developing countries are situated for the purpose of identifying the investment requirements of the petroleum industry and carrying out the relevant evaluation studies.   A second phase could be to set up machinery to finance activities, such as petroleum explorations, for which there is no access to existing sources of finance.   It would also seem desirable to establish within the financial institutions of the various regions a contingency fund which could be constituted with resources from both industrialized and developing countries.   This would undoubtedly be a factor in providing broader guarantees and in enabling state petroleum enterprises in the Third World to obtain adequate financing.

Because of the historical circumstances in which the world petroleum industry developed, financing by private organizations has a greater influence on the operations of petroleum enterprises.   The large transnational petroleum corporations have a pronounced influence on the banking and credit institutions in industrialized countries - so much so that, when state petroleum enterprises in Third World countries undertake oil development projects of indirect interest to the transnationals, the traditional barriers to obtaining credit are rather easily overcome.   Such projects generally involve increasing hydrocarbon production to supply the international market, or improving transport and marketing systems. The rate of interest is higher than that charged by international agencies and repayment terms are shorter. But banks and credit institutions sometimes allow more leeway for increasing domestic input into investment projects.   The large deposits of petrodollars and Eurodollars in private international banks in recent years have increased liquidity and, hence, their lending capacity.

The Mexican oil industry normally resorts to these credit systems, both for revolving credits and for medium-term financing.   These resources are channeled into investment projects with high profitability (mainly petrochemical plants) in order to minimize the risks resulting from escalating prices of machinery and equipment and from any change in the rate of exchange between the national currency and the foreign currency in which the loans are negotiated.

The issue of bonds by state petroleum enterprises in the stock markets of industrialized countries is not a common practice, in view of the private investor's lack of confidence in foreign enterprises. Petroleos Mexicanos has issued petroleum bonds fairly frequently on the United States market and in European countries; what has made this possible is the participation of consortia of banks and financial institutions. However, despite the advantages which might accrue from such methods of financing, negotiations will have to be intensified in the future to overcome lack of confidence on the part of investors.

In some cases, the credit provided to state petroleum enterprises by suppliers of goods and services is considered to be an important source of finance. There are various underlying causes for this:

1. In most developing countries the production of capital goods is in its infancy.

2. In investment projects for state petroleum enterprises there is a constant endeavor to use the most advanced technology in order to postpone obsolescence as long as possible.

3. The suppliers of goods and services in highly industrialized countries offer interest rates and repayment terms which clearly facilitate the execution of investment projects. The reason for this is that the sale of equipment and services by large transnational companies to state petroleum enterprises makes them captive customers for the future, since any change in technologies is practically impossible once projects financed in this way have been carried out.

4. Where the domestic industry does produce capital goods, prices are considerably higher than those of foreign suppliers, since the size of the market precludes economies of scale. Moreover, the credits which local producers are able to grant to domestic suppliers are at higher rates of interest than those provided by suppliers in industrialized countries, and the repayment terms offered are shorter. This is all because of the characteristics of the domestic money and capital markets.

In view of the foregoing, the development of state petroleum enterprises will depend on the characteristics of the financing granted by suppliers in the industrialized countries.

In Mexico, the technology of planning and design has developed rapidly, as have manufacture of some products for the petroleum industry and supply of specialized services. However, it has not proved feasible to compete, as was hoped, in the market created by the state petroleum enterprises of the region, despite the advantages of the transfer of technology that is offered, since the decisions taken by these enterprises have been determined by financing granted at rates of interest lower than those available in the international market.

The liquidity prevailing in insurance companies and pension funds in the highly developed countries enables them to provide financing for state petroleum enterprises through bond issues in view of the fact that interest rates in developing countries are normally higher than those available in industrialized countries. Petroleos Mexicanos has increasingly resorted to this source of financing, since the agreed repayment terms are longer and the rates of interest lower than those of foreign banking institutions.

In conclusion, there are a number of observations that can be made:

1. The setting of "policy prices" for hydrocarbons in developing countries prevents state petroleum enterprises from being financially able to meet the needs of the domestic energy market effectively. "Subsidized prices" for the hydrocarbons consumed in domestic markets are not a factor of any value in stimulating the economic development of countries; on the contrary, they

lead to undercapitalization of state petroleum enterprises encourage waste of energy, and represent transfers of state resources to the most favored social groups.

2. The prices of hydrocarbons in the international market and the pattern of the demand for energy in developing countries make it essential to have sources of finance and financial mechanisms conducive to the sound growth of state petroleum enterprises.

3. Domestic sources of finance are inadequate to meet the capital requirements of state petroleum enterprises.

4. The origin of the technology used in the petroleum industry and the rigidity of international financial mechanisms subject the investment projects of state petroleum enterprises to terms and policies laid down by the transnational oil companies. Thus, development of petroleum reserves of developing countries has tended to become dependent on the conclusion of "risk contracts," providing for large concessions to those companies of geological tracts possessing hydrocarbon potential.

5. Financing from foreign sources usually carries with it the obligation to purchase goods, services, and technologies from the countries granting the credits and, although in the initial stage the credit terms offered are attractive, state petroleum enterprises later become captive customers because of technological dependence.

6. Most of the financing obtained by state petroleum enterprises in Third World countries is in the form of "tied credits" as a result of which the necessary incentives have not been provided to stimulate development of domestic industries auxiliary to the petroleum industry.

In consideration of these points, there are a number of recommendations for state petroleum enterprise. It is necessary to establish domestic price systems for hydrocarbons based essentially on real production costs, plus a mark-up to allow financing of the additional investments required for the balanced growth of the state petroleum industry. International financial institutions should make their financing mechanisms more flexible and adapt them to the operational characteristics of the petroleum industry. International funds should be set up for preinvestment studies of the petroleum industry in Third World countries. The establishment of a contingency fund to guarantee against unforeseeable risks in the execution of Third World petroleum industry investment projects could broaden the horizon of international financing and liberalize the terms at which it is granted. State petroleum enterprises should seek machinery which will enable them to finance research for their technological development and reduce their external dependence. The formation of multinational enterprises to produce goods and services for the state petroleum and petrochemical industries in developing countries would make it possible to achieve economies of scale, as the enterprises would be large enough to satisfy the demand of two or more countries, and would increase the national input into petroleum and petrochemical investment projects.

15 Financial Provisions of State Petroleum Enterprises in Developing Countries

D.H.N. Alleyne

This paper examines the adequacy of the financial provisions made for state petroleum enterprises in developing countries in the context of the roles assigned to them. It argues that an adequate financial framework is a necessary condition for the success of the state petroleum enterprise; and that failure to provide this condition, which occurs often, not only hampers the growth, stability, and effectiveness of the enterprise, but can affect adversely the future development of the country itself.

Research has indicated that information on this subject is not generally available, that it has not been closely examined by economists or by most of the countries, and that above all, the fundamental problem springs from a rather unclear picture of the nature and the characteristics of the state petroleum/enterprise. It is important at the very outset, therefore, to delineate the concept of the state petroleum enterprise envisaged in this study. For the present purpose, a state petroleum enterprise is defined as an artificial, legal person created by act of the legislature or by government decree; an entity, that is a thing having real existence, and acting on its own volition; responsible, that is accountable, and able to sue or be sued in its own right; a corporation sole, or, less frequently, a corporation aggregate; self perpetuating, having a continuous existence irrespective of that of its members; autonomous, but not sovereign. In respect to these characteristics and objectives, the state petroleum enterprise operates within the general framework and guidelines established by its owner and creator, the state.

The past half century has seen increasingly rapid growth in the creation of state petroleum enterprises. By 1977, more than 80 state petroleum enterprises had been established in developing countries. This phenomenon has coincided with, and been accelerated by, the rapid increase in the number of independent states that have emerged from the status of colony or satellite of an industrial power. In some cases, the political change was gradual and relatively uneventful. In others, great trauma characterized the transition.

151

Going hand in hand with the movement for political autonomy has been the wish to own, control, manage, and exploit national resources for national ends. This has been given recognition by various UN resolutions, as for example, Resolution 523 (XI) of January 12, 1952 and Resolution 1803 (XVII) of December 14, 1962. This has gone hand in hand with attempts to reduce the dependence on the transnational petroleum corporations which have played such a predominant role in the development and control of the international petroleum industry.

In nearly every case, the state petroleum enterprise, with its subsidiaries, is seen as the chief instrument for increasing national revenues from petroleum and for both stimulating and satisfying the demand for oil and gas in the development of the economy.

Any entity attempting to break into the petroleum industry must take into account its unique set of characteristics. The industry is nearly always, especially at the exploration stage a high risk, aleatory, commercial venture. It is built around and developed through a special technology requiring constant updating. It concerns a flow commodity requiring an almost continuous stream from wellhead through refinery to tankage to pipeline or tanker to the consumer. Any blockage in the line immediately involves either the considerable capital expense of more storage facilities or the shutdown of production. It is capital intensive and demands large inputs of capital which may yield substantial profits but which, in view of the risk involved, may never be recovered. Its marketing sector, which may be the first or perhaps the only area of operations for many state petroleum enterprises, has come to be the area to which the transnational corporations, in organizing their integrated operation, have caused to accrue the least profits. (However, it must be noted that developments in the international petroleum industry since 1973 have set the stage for some changes in the earlier trend, so that more substantial margins may be restored to the refining and marketing sectors - to the benefit of the state petroleum enterprises and the transnationals in their new roles.) And where the transnationals have had to resort to external sources of finance, they have been able to do so from a position of strength based on their internal financial soundness, based on self-generated and retained cash.

It is proposed, therefore, to examine several state petroleum enterprises in developing countries in terms of their legal provisions and their procedural arrangements governing matters of finance and investment. Are they created with sufficient actual or potential financial strength to ensure that they grow and have a perpetual succession, or are they unwittingly launched into the world bound hand and foot in the hope that they may yet swim?

The following matters, particularly, should be examined as factors that separately and in various combinations affect corporate financial strength:

1. Authorized and paid up capital: Is the state petroleum enterprise provided with an adequate equity base for its operations? Who owns the equity?

2. Disposal of income and formation of reserves: What control is state petroleum enterprise permitted over the disposal of its income and profits? Is it permitted to follow proper business practices and reinvest, or does it have to pay all or most of its profits to the government?

3. Budgeting and investment planning: Does the state petroleum enterprise approve its own budget or is this controlled in detail from above?

4. Domestic pricing policies: Can the state petroleum enterprise charge economic prices at home, or is it constrained to charge politically regulated prices over all or most of the range of its product? What are the possible justifications for this? What are the effects on the company?

5. Loans: Is it necessary, on account of an inadequacy of equity, to borrow all its financial requirements? What are the implications of various debt/equity ratio situations? Must the state petroleum enterprise have permission to borrow at home or abroad? What international loan or credit sources are open to it, and what is the present situation in that market?

NATIONAL IRANIAN OIL COMPANY (NIOC)

The National Iranian Oil Company was established on April 29, 1951, following the nationalization of the petroleum industry, to manage, own, and operate the petroleum resources of Iran. It was not the first of the state petroleum enterprises established in developing countries, nor, indeed, was it the first state petroleum enterprise created in Iran. The Oil Company was established in 1949 as a public corporation to explore and prospect for oil. It is, however, among the most successful. Fortune magazine placed it second, after the Royal Dutch/Shell Group, in its August 1977 list of the 500 largest industrial corporations outside the United States ranked by sales.

NIOC Financial Data 1976 (Thousands of Dollars)

Sales	Assets	Net Income	Stockholders Equity	Number of Employees
19,671,064	6,544,991	17,175,182	4,261,182	57,331

The relevant enabling law provided for initial authorized capital (50 percent paid up) of 10 billion rials, divided into 10,000 registered shares of one million rials each, all nontransferrable and wholly owned by the government of Iran. (By 1975 this had increased to 100 billion rials.) The allocation of profits provided for a general reserve account to which the company is required to transfer every year two percent of its total income, and a special reserve to which appropriate amounts, as approved by the general meeting, are set aside from total revenue to meet the company's requirements according to good oil industry practice. Total revenue is described as the net profit plus payments as mentioned in the Oil Agreement approved on October 29, 1954; amounts gained, that is oil bonuses, land royalties, and refunds of exploration cost by virtue of the Petroleum Act of 1957; as well as sums received from subsidiary and affiliated companies, their net profit after deduction of applicable income tax.

It is clear to anyone who has had dealings with NIOC that it enjoys a certain esteem in Iranian government circles. It is the repository of the national will to own, control, and manage petroleum resources. It has, therefore, been entrusted not only with certain fixed assets, but has also been given an adequate equity base and ample scope for operating as a sophisticated corporation, to train its personnel and to plan, finance, and execute its programs. It also enjoys certain special privileges which enhance its reputation and assist in its operations. The strength of its financial base does not derive only from the legal provisions set out above. Those legal provisions, however, are a testimony of the relationship between government and state company and a necessary condition for its success.

CASE B - QATAR NATIONAL PETROLEUM COMPANY (QNPC)

Reference is next made to the Qatar National Petroleum Company (QNPC), not in order to make any direct comparison with NIOC, but because it is of more recent origin, somewhat smaller, and it possesses characteristics which other small producers may find interesting.

Law No. 13 of 1972 established QNPC as an independent legal company with an official seal and unlimited term. It provided for a share capital of 100 million Quatari rials to be wholly subscribed by the state in installments whose timing and size were to be decided by the board of directors in accordance with the requirements of the company and the approval of the Minister of Finance and Petroleum. It also directed the state to transfer and deliver to the company part of the capital as ownership of the net financial balance of the National Distribution Company and the state's share in the capital of the Qatar Fertilizer Company. The state was permitted to pay the remainder of the capital or any part of it in kind.

As regards the allocation of profits, the law provided that 10 percent of the net annual profits be set aside for the creation of a statutory reserve. It also provided that this appropriation should cease when the total of the reserve reaches 50 percent of the company's capital, and that the company pay annually to the state 55 percent of the remaining net profits, that is, 49.5 percent of the net profit. Finally, with the approval of the Minister of Finance and Petroleum, the company was permitted to retain the remaining profits (40.5 percent) as a general reserve for the purpose of expanding its operations.

CASE C - IRAQI NATIONAL OIL COMPANY (INOC)

In the case of the Iraqi National Oil Company (INOC), the law provided that 25 million Iraqi dinars of share capital be wholly subscribed by the government upon the request of the board and with the approval of the Council of Ministers. (Other sources give the total authorized capital of INOC as I.D. 70 million in 1977.) Any part of the unpaid capital was to be considered as guanranteed by the Iraqi treasury until such time as the whole of the authorized capital was paid up. The limit for the authorized capital could be raised to a maximum of 150 million dinars - if proposed by the board and approved by the Council of Ministers. The assets of the company included the grant by the government, without recompense, of oil industry installations pertaining to any field or part of a field in the areas assigned to the company for development. (It is interesting to note that in some countries the concept of the relationship between state and state corporation is such, and so great is the obsession with safeguarding the treasury, that sometimes it is seriously questioned how much and when the wholly-owned government corporation will pay its sole shareholder, the government, the money-value of the assets vested in the corporation. In fact, such vesting of assets is often the best way to put them to work for the eventual benefit of the treasury and the economy as a whole.)

Whereas, in the case of NIOC, 50 percent of the authorized capital is paid-up; and with QNPC, the capital is to be subscribed by installments as and when required; the provisions in respect to INOC do not set out the initial amount of paid-up capital. Each country, affected as it is by different factors, tailors its legislation and its practices to suit both its corporate pocket and its overall domestic circumstances. To meet the "defect" of having no initial paid-up capital, the INOC statue provides that such part of the authorized capital as remains not paid up is considered as guaranteed by the Iraqi treasury. This puts the total resources of the state up to the limit of the authorized capital as backing and support for the company. It can be argued that since the company is state-owned, such backing would have been obvious in any case. However, the fact of its being specifically provided in the law gives it that much more status and acceptability.

It is clear from the way that profit allocation was specified that the founders of INOC recognized the need to remedy the shortcoming in the initial provisions for equity capital and sought to fill the gap. This was done by providing that the company's net profits should be added to its paid-up capital until such time as the authorized capital was fully covered. It was also stipulated that after five years, starting from the time a net profit was realized, the company would pay 50 percent to the government until such time as the authorized capital was fully covered.

Thereafter, the company would pay 75 percent of its net profit to the government and the balance into its reserve fund (provided that the balance was at least 5 million dinars), until such time as the reserve fund reached the amount of the authorized capital. It was further provided that when the reserve fund reached four times the amount of the company's authorized capital, the company would pay all of its net profits to the government.

However, the effectiveness of these savings provisions were necessarily dependent on the ability of the company to earn income and profits in the first place. This ability is influenced by the nature of the assets assigned to the company, the promptness with which they could be made to earn an income, and by the company's ability to finance operations in the interim. This method of financing is not likely to be applicable and effective in the case of the state petroleum enterprises of those developing countries that do not yet have or control reasonably productive petroleum resources (in the short run) or do not otherwise have financial resources that would engender confidence on the part of dependable prospective creditors or partners.

CASE D - THE TRINIDAD AND TOBAGO NATIONAL PETROLEUM COMPANY (TTNPC)

The Act of 1969, under which The Trinidad and Tobago National Petroleum Company was established, provided no authorized capital for the company. It provided, however, that the assets of the company would consist of such exploration, exploitation, and related rights as may be assigned to it by the government, any real and personal property as may be transferred to it by the government or acquired otherwise, and that portion of the profits which it may hold for the development of its activities.

There was, however, no given schedule of rights, assets, or properties actually assigned or transferred which the company could develop or exploit. It is necessary, therefore, to examine the provision for the allocation of profits to see whether the formal provisions attempt, as in the case of INOC, to fill the gap. As to allocation TTNPC's profits, it was provided that the company should conduct its activities along business lines and that any profit realized by its operations should accrue to the government. Such proportions of these profits should be used for the purpose of expanding the activities of the company and for the provision of special reserves or of sinking funds. In any one year, the government could allocate to the company the whole or part of any sum required to cover any deficit disclosed by the budget.

It is for consideration whether the provision that the company should conduct its activities along business lines is consonant with the latter part of that same provision which states that all the profits of the company should accrue to the government, or with the procedures involved in the provision determining the disposition of profits. But times and circumstances often dictate a course of action, and many small, producing or nonproducing countries may find themselves persuaded that the needs of the national treasury are so great that it must compromise with respect to what may be considered ideal financial provisions for the creation of a state petroleum enterprise. The question is whether the success of the enterprise is, thereby, assured or compromised. The TTNCP has had a board of directors appointed. Beyond this, it exists only in the statute. Other state petroleum enterprises have, however, been created, including the National Petroleum Marketing Company and the Trinidad and Tobago Oil Company, both of which may logically have been subsidiaries of the TTNPC.

CASE E - THE NIGERIAN NATIONAL OIL COMPANY (NNOC)

One should next look at one of the state petroleum enterprises in Africa, south of the Sahara. Nigeria, by a Decree of April 1, 1971, established the Nigerian National Oil Company as a statutory corporation with perpetual succession and a common seal. It made, however, no specific provision for authorized or paid-up capital or the assets of the corporation. It provided that the commissioner, with the approval of the Federal Executive Council, could issue to NNOC such directions as he might think necessary as to the disposal of any surplus funds of NNOC and, subject to any such directions, NNOC could invest its funds and maintain a general reserve.

NNOC was directed to maintain a fund consisting of grants or loans provided by the Federal Executive Council and of monies received by it in the course of its operations. This fund was to be used to defray all expenses incurred. The corporation was directed to submit, not later than three months before the end of each fiscal year, estimates of its expenditure and income (excluding payments to the corporation out of monies provided by the Federal Military Government) Relating to the next following fiscal year.

But the NNOC of 1971 has ceased to exist. The company and the Ministry of Petroleum Resources were merged by Decree No. 33 of 1977, which also created a new single entity, the Nigerian National Petroleum Corporation (NNPC), to replace them. The decision to merge, it seems, was motivated by the belief that bringing the public sector of the oil industry under unified control would lead to better understanding and cooperation among its staff and promote the optimum utilization of the scarce indigenous human resources available in the Nigerian oil industry.

There was a further change in that the new corporation (NNPC) reports directly to the government instead of through a ministry, as was the case with NNOC. As regards the financial framework, there has been no important change in the earlier concept of the national petroleum corporation. The financial arrangement continues to be determined by the view that "as a corporation wholly owned by the Government, the NNPC has no fixed capital, but is dependent on budgetary grants from the Government." (1)

It must be the case that each country, at any point in time, makes its own judgments and its own decisions about its affairs. In fact, such judgments may be the best in all circumstances, especially since these circumstances may not be known or properly appreciated in their entirety by any external observer. It seems appropriate, however, to make a number of observations. The functions of a ministry of petroleum resources and the attitudes and modes of operation of its staff are quite different from those of a state petroleum enterprise. A ministry, or government department, and a corporation are distinctly different institutions conceptually and functionally. It is doubtful that this new entity will be able to satisfy both. Given the other provisions for direct contact with the central government and for funding the corporation, the most likely outcome will be that ministerial functions will be satisfied at the expense of corporation concerns.

The many examples of viable state petroleum enterprises wholly owned by governments indicate that such corporations need not be deprived of their essential equity base nor is it essential that they be dependent on budgetary grants from the government. The fact is that the institution thus created is not a corporation at all. It is, at best, a commercialized government department or ministry of petroleum resources. It is possible that an entity with this type of financial provisions may experience no delay in having its budget approved and in receiving the necessary grants for the implementation of its policies and programs. In fact, the NNPC received for 1977-78 an operating budget of $5,405,000 and for the period 1975 to 80, $7.5 billion for capital projects.

From time to time an enterprise may be short of cash or need infusions of new capital. Conducting its business of the basis of annual grants permits it only to exist. It can take root, flourish, and establish an independent identity only on the basis of its own capital. Its attitude towards a grant is necessarily different from its attitude toward its own capital. The latter must be invested and managed to produce the profits that are the basis of its self-perpetuation and the financial reward of its owners, the government. The corporation can take calculated risks with its own capital, risks that will instill a new spirit and open new avenues of enterprise and training for its personnel. And it is this personnel who will manage its affairs in the future, and move into other areas of national economic activity. It is hoped that this new institution in Nigeria is but an interim measure necessitated by specific local circumstances and that it will, in due course, give way to a properly constituted, financed, and managed state petroleum enterprise that is responsive to the needs of the country and the policies of its government.

Developing countries must consider the problems that ensue when a company has to prepare its budget with a substantial part of its revenue unknown, and when capital and recurrent expenditure items may be trimmed or excised by persons and for reasons external to the company. Such constraints may render it weak, ungainly and badly coordinated. Whatever may be the merits of this type of budgeting for a government department, it cannot be effective in an industrial corporation in a competitive situation.

It is worthwhile to recall the provision in the legislation establishing the Iraqi National Oil Company specifying that company should not follow the financial and accounting principles applied in the government. As regards INOC, the legal safeguard is double-barrelled because it was also specified that strict commercial accounting principles should be used.

CASE F - THE KUWAIT NATIONAL PETROLEUM COMPANY (KNPC)

Consideration of The Kuwait National Petroleum Company (KNPC) introduces an important variation. Whereas all the enterprises so far examined have been wholly state owned corporations, the KNPC is a mixed-venture company. It was established on October 3, 1960 under the law governing commercial companies, and is owned 60 percent by the state and 40 percent by citizens of Kuwait.

The initial share capital was 100 million Indian rupees. Only Kuwaitis were permitted to hold shares, and no member of the public was permitted to hold more than 3,000 shares by subscription or more than 6,000 shares by a combination of subscription or transfer except where the excess shares were obtained by inheritance. Profits were to be allocated in a manner similar to that of an independent company. Deductions from gross profits were to be made for depreciation and replacement of equipment and other assets. Net profits were to be allocated 10 percent to a compulsory reserve account and an amount to be determined was to be set aside yearly as a general reserve. That part of the profits to be specified by the general meeting were to be reserved for meeting company obligations resulting from the labor laws, and the balance was to be distributed in proportion to the number of shares.

KNPC is an interesting departure from the general pattern of wholly-owned state petroleum enterprises. The objective, apparently, was not only to have the company locally owned and controlled, but also to give nationals of the country an opportunity to participate in ownership of the company. However, it may be remarked for the benefit of developing countries wishing to open these opportunities to their citizens that the upper limits for individual shareholding were, in this case, somewhat generous.

To take a hypothetical case, if all applicants were given the maximum number of shares permitted, then only 266 citizens could participate. If resale took place within that group, the number of shareholding citizens could be reduced to 133. By application of the inheritance clause, the concentration of ownership would be even further accentuated. It may be preferable to include an arrangement for all applicants to receive a minimum number of shares on a first-come, first-served basis.

Unlike the other state enterprises, the KNPC was not given any state-owned assets, although it was given a monopoly of the local market for petroleum products. This was at a time when transnationals operated concessions in the country. There is, however, a growing trend in developing countries, not least in Kuwait, towards complete state ownership, control, and management of national petroleum resources.

For those developing countries that contemplate this interesting corporate structure, it should be pointed out that one has to overcome the problem of vesting the national assets in state enterprises without discriminating against those citizens who do not own original shares in the company. It may be that the name, composition, ownership base, or scope of operation of the company would have to be adjusted. It may even be necessary to create a completely new and different state petroleum enterprise to meet this new situation. In fact, this is precisely what Kuwait has done. (2)

CASE G - YACIMENTOS PETROLIFEROS FISCALE DE ARGENTINA (YPF)

In June 1922 the government of Argentina, faced with the problem of ensuring petroleum supplies for its armed forces, created The Yacimentos Petroliferos Fiscale (YPF), the first state petroleum enterprise in Latin America and the first in a developing country. Given the international situation at the time, this was a very courageous step. It is true that in 1917 Mexico amended its constitution, making it the earliest of the radical constitutions of the twentieth century. This had a major impact outside of Mexico, especially because it declared not only that private ownership was subordinate to public interest, but also that subsoil wealth was inalienable national property. It was not until 1938, however, that the assets of foreign petroleum companies were nationalized in Mexico, and vested in the newly created state petroleum enterprise, Petroleos Mexicanos (PEMEX).

Argentina took the initiative at a time when the nineteenth century attitude of the major powers towards the unindustrialized countries had changed but little. It is true that George Canning, the British statesman (foreign secretary and then, briefly, prime minister), had indeed prevented European intervention in South America on behalf of Ferdinand VII of Spain, and that he had recognized the independence of the rebellious Spanish-American colonies. However, it was the same Canning who, in reference to economic control, remarked, "Spanish-America is free, and if we do not mismanage our affairs sadly, she is British." There were other industrialized nations that shared similar views of Latin America.

Given this obviously hostile environment, what were the financial provisions made to sustain Argentina's bold venture? Harry O'Connor, in his book World Crisis in Oil, states that "YPF received initially an appropriation of 8,655,000 pesos (less than $4 million) and the Government never added another penny in all the subsequent years" to 1963. (3) By any standards, this must be regarded as a trifling financial commitment, given the task facing the company.

A review of the history of the oil industry in Argentina indicates that financial inadequacy undermined the capacity of YPF to perform effectively and to advance towards the national goal of petroleum self-sufficiency. This financial weakness opened the door to political infighting and to foreign pressure upon successive regimes to reverse the national policy, minimize the role of the YPF, and reintroduce private oil concessions. In 1950, YPF was deprived of its autonomy and financial independence. Hanson, writing in 1960, observed that "the Government had signed away proven reserves to certain foreign interests...on terms so prejudicial to Argentina in the light of prevailing industry practice that many veteran oilmen were astounded...." (4) Referring to the YPF exploration program drawn up in February 1966, under which the oil companies were limited to the role of subcontractors, Tanzer states, "However, financing such a programme was the critical problem, and the Illia Government appeared to be desperately grabbing at straws for a solution," possibly through a World Bank loan. (5)

In 1967, Argentina was once more opened up to foreign companies - the relevant law giving certain tax concessions and providing a tax rate reduction over a ten-year period for companies taking quick advantage of the new rights. There was a minimum tax of 55 percent, including royalties and rentals on net income, a retrograde step in the light of practices in other developing countries. Even in 1962, in a report to the UN, Walter Levy remarked that "while arrangements such as in Argentina have proved effective, we do not suggest that they are in any sense models that should be copied elsewhere." (6) This was the result of the underlying weakness originating in a persistent financial inadequacy.

CASE H - PERMINA

When Indonesia established Permina, one of the three state enterprises that would subsequently be united to form Pertamina, the country was faced with an industry that had been ravaged by war, revolution, domestic strife, and neglect. Capital was needed for equipment and materials for training managerial and technical staff. The company had to depend on foreign investment because Permina could expect no funds at all from the Indonesian government. Even so, the principal objective always was that Permina, as a national company developing a national resource, was to maintain control and management.

It has been demonstrated that some of the problems experienced by Pertamina in recent years were the result of injudicious expansion into certain fields of economic activity. Attention has been drawn particularly to the arrangements for the tanker fleet. It is equally clear, however, that the relatively weak financial base of the company was the initial problem. This led to the introduction of operating companies under conditions in which "returns to the operating companies are higher in Indonesia than in other producing countries, but so are unit capital costs." (7)

One of the factors that led to Pertamina's financial crisis was "a decline in foreign exchange earnings from crude oil." (8) This was at a time when crude prices were climbing rapidly. However, it was also seen that

> Indonesia's future role as an energy exporter is tied to its evolving relationship with the energy industry. The companies will seek a stable investment climate and further financial incentives. For its part, the government may have to make further modifications in the terms of its production sharing and work contracts. The work and production sharing contracts satisfy Indonesia's political requirement for ownership of its national resources. Indonesia also conserves its scarce capital and technical resources and is freed from the risks of exploration and development. (9)

While there is obvious room for cooperation between the transnational corporations and state petroleum enterprises, one wonders whether the price of such joint efforts, in a situation where the state petroleum enterprise suffers from a basic financial disability, may not be excessive. Another question is whether the developing country is not, thereby, left with the outer trappings of political sovereignty and ownership of natural resources while being pressured into conceding to others the real substance of the gains from the exploitation of those resources.

Under normal operating conditions, a company will charge for its products a price which permits it to cover all costs and provide at least a normal profit. Some may say that a petroleum company should charge what the traffic will bear. A state petroleum enterprise, however, operating within the domestic market, usually finds that it has motives and objectives somewhat different from those of the transnational petroleum corporation. It has to be a genuine corporate citizen, paying particular attention to the public welfare. It cannot unilaterally take advantage of any monopoly position that it may enjoy in order to achieve a high rate of return on investment by charging high prices on its full range of products. This is not the conduct or policy that the government expects. Even more important, this is not what the public expects. Because the state petroleum enterprise is more or less part of the government, a responsible conduct is expected of it. It is open to all manner of criticism.

At times of rapidly escalating price levels, a government may require the state petroleum enterprise to charge low domestic prices with the stated purpose of restraining the inflationary trend. It is not always carefully analyzed whether such price restraints in petroleum products do, in fact, have the anticipated effect. In areas such as public transport and electric power generation, the lower factor cost can have a moderating influence on the final cost of the service with beneficial effects on the economy. In such cases the requirement to charge political prices can be justified.

There are other examples. In one country, kerosene and LPG prices were set at levels to provide relatively cheap fuel in the hope of halting, and even reversing, the deforestation of the countryside brought about by the demand for firewood. The success of this policy promised considerable benefit for the weather, climate, agriculture, water supplies, soil and so on.   In certain other cases, however, specifically gasoline prices, the expected benefits are not always realized. There have been cases where the lower product prices lead to extravagance and wasteful use of the commodity.   Whatever the arguments and justification for or against political pricing, the fact is that the establishment of such prices is the prerogative of the government, which alone can make the political decisions required in such a case.

It is necessary, however, to examine the possible effects on the state petroleum enterprise of political pricing. Where the domestic market is small in relation to the total operations of the company, very little harm may be done. However, many of the smaller, recently created state petroleum enterprises are, of necessity, either confined to domestic marketing or to small-scale refining based on purchased imported crude plus domestic marketing. They may have to arrange for disposal of those items from the product slate that are not generally demanded in the domestic market. These state petroleum enterprises are likely to find that, even with normal operations and pricing, there are only small margins of profit at the refining and domestic marketing ends. They may find good spot markets for other export products from time to time, but on the whole they are going to be targets for clever operators who attempt to exploit their vulnerable position.   If, in addition to that disability, the state petroleum enterprise is required to charge political prices at home, the financial position of the company is completely undermined.   It ceases to be a profit-oriented, self-perpetuating entity, and, instead, becomes little more than a glorified service department or a species of public utility.

But domestic pricing   policies are important for the larger state petroleum enterprise as well.   The case of Italy's Ente Nazionale Idrocarburi (ENI), and the course of action it took successfully to safeguard its interests in a situation where nearly all fuels in Italy were subject to price controls, is discussed later. In the case of PEMEX, it would appear that while the national economy may have received a boost from lower prices for petroleum products, there were the negative results of wasteful consumption and some financial embarrassment for the company. "Undoubtedly an important factor leading to this rapid growth in petroleum demand was the government policy of keeping petroleum prices very low (a cause of much of PEMEX' financial problems)." (10)

Thus far, it has been demonstrated that some state petroleum enterprises have an ample equity base and control over disposal of earnings, as is usual with a corporation. Some even have statutory obligations to build up reserve funds. Some state petroleum enterprises, while lacking a stated equity provision, have been provided or promised financial inputs from their governments. The Petromin statute, for example, provided that its assets would consist of funds contributed by the state treasury and possibly an advance from the Saudi Arabian Monetary Agency.

It is hardly likely that the state petroleum enterprise of a wealthy oil producing country - particularly the world's largest crude exporter -

would be left short of funds. The company to which the Aramco assets, now nationalized, may be assigned, has the potential to become the largest and strongest of the state petroleum enterprises. One wonders, therefore, whether a more formal framework making such a company a fully autonomous legal entity, adequately financed and operating within the general control of the state, would not be a necessary prerequisite for the important role which a Saudi Arabian state petroleum enterprise will play in the international petroleum industry.

A number of state petroleum enterprises, however, are in the unfortunate position of being required to operate without an equity base, sometimes without the formal assignment or vesting of production assets, and without any control over the proceeds from their operations. The result is that any profits which they earn accrue to their governments. Further, they do not control the formation of their reserves as a corporation. These state petroleum enterprises may receive financial inputs from their governments, but these are unpredictable, as both timing and amount.

The other avenue or source of finance that may be available to state petroleum enterprises is loan funding. This borrowing can be from domestic sources where the strength of the local economy so permits. However, in most developing countries, only relatively small sums can be raised internally, although such borrowing can be helpful in that it may be a stimulus to the creation or growth of a local stock and money market. On the other hand, it may attract to the petroleum industry funds that might otherwise have gone to small enterprises that are less likely to attract funds from the international market.

In any event, where a large part of a loan is required for the acquisition of foreign goods and services, it must be either denominated in one or more foreign currencies or be connectable. In this case, it is an actual or potential drain on the foreign exchange resources of the country. To protect the economy and the value of the national currency, it is essential that foreign borrowing be controlled or at least monitored by the central government.

Some countries permit domestic borrowing by their state petroleum enterprises with no restriction or prior permission, but require specific application and approval for all foreign borrowing. Some do not permit their state petroleum enterprises any borrowing whatsoever without the prior approval of the government. This restriction is often associated with state petroleum enterprises that have no equity base, no specific assets, and no company control over the proceeds of the operation.

The provisions and procedures that state petroleum enterprises have to observe on borrowing vary widely. NIOC is permitted by its statute to obtain and grant credits and loans within the country or abroad for the purpose of achieving the company's objectives. The granting and obtaining of loans and credits from abroad are, however, subject to approval of the general meeting of shareholders' representatives. But domestic credits and loans can be made by the company without such prior approval. In Nigeria, NNPC does not, except with the general or specific approval of the Commissioner, have the power to borrow money or to dispose of any property, although there is a possibility of loans and grants from the Federal Executive Council. Petromin is permitted to contract loans, accept gifts and donations, and own movables and immovables. The TTNPC may borrow temporarily by way of overdraft or otherwise such sums as may be required to meet its obligations or discharge its functions. Additionally, subject to the approval of the Minister of Finance, the company may borrow funds for working capital or for the acquisition of shares or other interests in companies engaged in petroleum or petrochemical industries; the establishment subsidiaries; for meeting expenditure chargeable to the capital account, including the repayment of any money borrowed by the company for defraying such expenditures; and for similar purposes. The Minister of Finance may guarantee (with or without conditions) repayment of loans raised by the company. Iraq's INOC can secure loans or credits inside or outside Iraq to finance its projects, but external loans require the approval of the Council of Ministers.

A state petroleum enterprise of a financially strong developing country is unlikely to find itself in desperate need of money. However, even the more powerful and successful enterprises may have to resort to borrowing when undertaking large projects. Those that are financially sound may have no problem in raising loans. However, where there is no equity base, no control over earnings, and where no assets are vested in the company, or where such assets as may have been vested in the company have no early prospects of earning substantial income, problems do arise. The balance sheet and prospectus that such an ill-financed state petroleum enterprise can provide to a prospective creditor may not be sufficient. The most likely outcome is that the state petroleum enterprise will not be able to secure a loan in its own right. It will have to depend on its government either to borrow on its behalf or to guarantee the loan. Furthermore, financial inadequacy does not long remain a secret, and can hamper the state petroleum enterprise in every other area of industrial or commercial activity. It will undermine its bargaining power and reduce its leverage in areas where it otherwise would have had no problems.

It may be argued that it is not a serious matter if a state petroleum enterprise requires its government to secure loans for it. Such an argument, however, overlooks the fact that an institution whose scope for action is thus reduced and circumscribed by its dependency on the state falls short of the concept of an enterprise that is an independent organization. It ceases to be an autonomous legal person or an entity responsible for its actions and its self-perpetuation.

Further, this argument does not take account of the fact that while such a state petroleum enterprise may have no financial difficulties as long as the coffers of the central government are full, that is precisely the time when the state petroleum enterprise should stand on its own against the day when the rest of the economy may be unable to continue to support it. At that time, the state petroleum enterprise may have to make a major contribution to national economic stability in an inhospitable international environment.

There is, however, considerable variation in the prospects for borrowing by developing countries because of the wide range of stages of development, ownership, and exploitation of petroleum resources, international political affiliations, and so on. One can discern three broad categories:

1. The powerful oil-producing and - exporting countries of the developing world have fully integrated and consolidated state petroleum enterprises that have little difficulty in raising loans. In fact, they find prospective creditors beating a path to their doors.

2. At the other extreme are the poor developing countries. They have a narrow and weak economic base, are not oil producers, and own state petroleum enterprises that often have no more than a stake in their small domestic petroleum marketing sector. The prospects of such governments securing loans on the international market are much less bright.

3. In between these two extremes are countries at various stages of economic growth and development. Their state petroleum enterprises have a greater or lesser degree of integration and may be operating in two or more sectors of the petroleum industry.

The countries in the first category are few and far between. The large numbers are in the second and third categories. Those in the second category represent a special case which should not be dealt with in terms of the competitive international market. Instead, they should receive special treatment and assistance from both the developed countries and the relatively prosperous developing countries. Those in the third category represent the real challenge. They do not have the resource of the first category nor the limited opportunities for action of those in the second category. They have to act and commit themselves.

It has been possible in the recent past for the governments of the countries in the third category to raise quite substantial loans through international or private banking institutions and suppliers credit, and thereby have been able to facilitate petroleum investment. But this situation will not necessarily persist. A number of factors threaten the continuation of this ease of borrowing.

It has been stated that after the OPEC price increases of 1973 and the consequent accumulation of foreign exchange reserves by the oil exporting countries, there was a panic among the major Western industrialized nations to halt this growth of economic power. According to Robert Engler:

> There was also frank fear among industrial powers such as the United States that OPEC members would press for greater voting power in the IMF and World Bank commensurate with their new wealth and contributions. Such representation would weaken the Western hold over these agencies and the ability to shape modernization in the capitalist mold including financial discouragement of Third World countries from developing petroleum and other resources under national rather than corporate agencies. (11)

That the situation must be taken seriously is further emphasized by the recent report of a United States Senate Sub-Committee entitled: "International Debt, the Banks and U.S. Foreign Policy," which has given rise to serious questioning of the growing debt of the developing countries, especially to the United States; the problems faced by these countries in meeting their debt obligations; the fact that developing countries need to spend between 20 and 40 percent of their export earnings to service their external debt; and the further estimate that by 1980, 25 to 50 percent of all new borrowing by developing countries will go towards repaying old debts.

Certain countries such as Brazil, Mexico, Argentina, and India, which have been making tremendous efforts to industrialize and have given prominent roles to their national petroleum enterprises in the development process, have been singled out for special mention.

The New York Times of September 18, 1977, in commenting on the report stated, "Brazil for one, with an external debt of about $25 billion to $30 billion, will have a debt servicing burden this year of nearly $5.3 billion or 40 percent of an estimated $12 billion in export earnings. And even with the 400 percent increase in the price of coffee, the study says, Brazil may have to borrow an additional $6 billion to finance its 1977 payments deficit."

One cannot, therefore, be complacent about nor presume upon the continuing ability of all developing countries and their state petroleum enterprises to borrow for investment. UNCTAD has realized the gravity of the problem and in its discussions of the New International Economic Order has asked for a waiving of loan repayment. Having referred to the "heavy debt service payments, current account deficit, etc..." UNCTAD IV recommended in February 1976 that "debt refief should be provided by bilateral creditors and donors in the form of waiver or postponement of interest payments and/or amortization cancellations of principal etc. of official debt to developing countries seeking relief." (12)

It is clear, therefore, that loan financing may not be generally as available to state petroleum enterprises directly or through their governments, as was once the case, and that there may, in fact, be no substitute for properly organized state petroleum enterprises with adequate financial provisions supporting a framework for effective operations.

So far, this examination has been confined to state petroleum enterprises in developing countries. It is proposed, however, to discuss one state petroleum enterprise of a developed industrialized nation, Ente Nazionale Idrocarburi (ENI) of Italy. A study of ENI is interesting for developing countries, if only because it indicates the steps taken by a small developed state to achieve self-sufficiency in energy supplies. One can also observe the flexibility with which the corporate structure was employed to attain certain sociopolitical and economic objectives which are, by and large, the same as those of the state petroleum enterprises of the developing countries. ENI introduced a number of new concepts into agreements and arrangements with developing countries that facilitated the creation and viability of their national petroleum enterprises. ENI is now the holding company for a vast petroleum, industrial, and commercial empire. It is, however, one of those strange institutions that are born or created after their offspring.

Italy first attempted to find oil and gas by providing subsidies to wildcatters on a per meter basis. When this did not achieve results, Azienda Generale Italiana Petroli (AGIP), now a subsidiary of ENI, was created in 1926 under company law with stock owned largely by public agencies. Its purpose was to engage in petroleum research and commerce. Those developing countries that are discouraged by the early difficulties of the state petroleum enterprise may take heart from the fact that the only thing that permitted AGIP to continue in active existence and operation was the stubbornness and optimism of Enrico Mattei who, in May 1945, disobeyed instructions to terminate AGIP's exploratory work.

ENI was created on February 10, 1953 as a holding company "with a clean charter for an active and dynamic business operation in the interests of the nation." (13) It is today listed as number six in the Fortune directory of the 500 largest industrial corporations outside the U.S. (14) (ranked by sales), with a range of activities and group of subsidiaries that span the petroleum, petrochemical, and related manufacturing industries.

In what may appear to be a position in conflict with the earlier argument against an excessive dependence on loan capital for investment and the implied suggestions about the dangers and problems of having too high a debt/equity ratio in the young state petroleum enterprise, attention is drawn to the following observation: ENI has an extraordinarily high ratio of indebtedness to physical assets. To finance a forced draft expansion aimed at resource control it appears that ENI uses increases in net income not for the direct requisition of assets, but to cover the interest charges and amortization of new capital over a term of years. In this way growth may be four or five times as great as would by the case if new investment were financed entirely out of earnings. (15)

But the ENI example really strengthens the case for an adequate equity base and for company freedom based upon sound business practices. It points up the need for at least one but, if possible, several areas of substantial profit earnings (even within the domestic market), and for using those earnings for servicing whatever debts may have been contracted, while at the same time reinvesting profits to promote the growth of the entity. All of this has taken place within the context of strong government support and encouragement for the corporation, which acts in accordance with clear national guidelines and objectives.

It was observed that certain state petroleum enterprises have no control over the utilization of their income. This is not the case with ENI, as "financing the various ENI operating groups has been effected largely through a policy of almost complete reinvestment of profits." (16) From 1957 to 1960, it was estimated that self-financing would cover 71 percent of the investment program. But, unlike many other state petroleum enterprises, ENI was able to invest and borrow on both the strength of an adequate equity base and satisfactory powers for creation of reserves:

The law of 1953 creating ENI assigned to it a capital of 30 billion lire, approximately 15 billion as the nominal value of the portfolio assigned to the company, and the remainder from the Treasury. ENI's net profits are by statute assigned 65 percent to the State; 20 percent to reserves and 15 percent to research and the preparation of personnel for work in the industry. The State's quota for the first three years was assigned to ENI. As a matter of policy, ENI's net profits (as a holding company) are essentially nominal. They revolve around 15 percent of its original capitalization for a number of years. It seems quite clear that ENI is considered as an instrument of capital accumulation in the interest of the polity, not for its profits. (17)

On the question of pricing and earnings in the domestic market, an area where some state petroleum enterprises are frustrated by what appears to be the excessive use of political pricing, the experience of ENI may also be useful. In Italy, most fuels except natural gas came under price control. ENI, therefore, paid particular attention to this potential area of profit in both the production and distribution of natural gas. As Dechert remarks: "This highly profitable operation was used to finance the overall development of the ENI complex. ENI set a price for natural gas competitive on a caloric basis with fuel oil. As a result a substantial sum has consistently been available for investment and expansion, probably abut 20-25 billion lire in 1960." (18)

It should not be concluded that only such large state petroleum enterprises as ENI or NIOC are cognizant of the benefits of corporate strength, investment on the basis of self-generated cash, strong equity base, and a reasonable debt/equity ratio. CEPE of Ecuador reported in 1977 that with respect to its "extensive investment program in the field of exploration, exploitation, refining and development of basic infrastructure for marketing the hydrocarbons to the best advantage of the nation...approximately 60 percent of the total investment has been financed by CEPE's own resources." (19)

It may be appropriate to close this section with a further exhortation for developing countries and their state petroleum enterprises to pursue this objective of an adequate equity base and a debt/equity ratio at least in the region of 80/20. In this connection, one can do no less than observe that the law establishing INOC provided that the aggregate of the company's outstanding loans must not amount to more than four times its authorized capital.

What emerges from this study is the wide variety and complexity of forms and systems governing financial provisions for the various national petroleum enterprises in developing countries. At one extreme there is the state petroleum enterprise that enjoys the full confidence of the country. It is provided with an adequate share of capital and productive assets so that it can immediately exploit national petroleum resources, regulate its own finances, and plan the training and development of nationals to both manage the affairs of the company and engage in research. Such a state petroleum enterprise, as its operations expand and as additional capital funds are required, has its authorized and paid-up capital and its reserve funds correspondingly increased. In this way, not only are the interests of the company served, but human and other resources are prepared for possible assistance or transfer to other sectors of the nation in the interest of overall national development. However, not all state petroleum enterprises that have been assigned potentially productive petroleum areas have made optimum use of their opportunities. Very often, this failure can be traced to restrictions on autonomy and on financial inadequacy.

At the other extreme, one sees the state petroleum enterprise which is unprepared for the role it was ostensibly created to fill. This type of corporation is almost doomed to failure. Some never achieve an existence beyond the drawing board or the law that created them. Some remain, while others have their original roles reduced and part of their functions transferred to the very transnational petroleum companies that they were supposed to replace.

Every weak state petroleum enterprise should be examined in terms of its financial position and the concept of independent operation. Some state petroleum enterprises are weak because there was no clear concept of the nature of their role or of the characteristics required for success in the international petroleum industry.

Some were created hurriedly to meet an urgent situation without proper consideration of all the implications. In some cases, the desire to create the state petroleum enterprise was not matched by the financial resources of the central government. Yet in others, there was an irrational and wholly unjustified concern to insulate the treasury from current or prospective financial loss or contingent liabilities. Though the treasury may have been quite healthy, an inherent conservatism may have prevailed over other political, economic, and social justifications for a strong state petroleum enterprise.

Again, the state petroleum enterprise may have been deliberately kept weak because of the reluctance of the central government to create an autonomous entity with the basic ingredients for growth and strength lest it become a challenger for the limelight of political power and influence. Despite the wish to create a strong institution to challenge the transnational corporations and localize control and management of the petroleum sector, there is, at the same time, a reluctance to permit a new power to exist and operate independent of the central government. This ambivalence results in action that keeps the state petroleum enterprise financially weak and without any real decision making power. This forces it to come to the central government for investment, working capital, loan approvals, and for detailed examination and approval of its budget.

Further, failure to meet national expectations is often seen as an occasion for panic rather than the time for infusion of required capital, and is often dealt with by increasing the financial limitations and administrative control on the part of the central government. But there is no way that a strong central government can be thwarted by a state petroleum enterprise. The state petroleum enterprise is a subsidiary to the government. To be effective it must be autonomous, but it is not and cannot be sovereign. In an extreme situation, a government can easily remedy any threats. In most cases, it is possible, although perhaps not always easy, for the central government to ensure competence, dedication, and loyalty through careful selection of the chairman and members of the board of directors and the managing director of the enterprise.

There ought to be such a close relationship between the top management of the state petroleum enterprise and the head of the government or the ministers directly responsible for finance and petroleum that no important decisions of the state petroleum enterprise are made without the full knowledge of the social and economic fabric that every decision must be examined in terms of its implication for the nation. Any state petroleum enterprise's chief executive who does not recognize this fact together with the concomitant need to consult, inform, and discuss his actions with the government is not worthy to head that organization. At the same time, the government, having enunciated clearly its national objectives and policies, must recognize the need to leave the enterprise free to operate within that framework. There must be frequent meetings between ministerial and company heads where information and views are constantly exchanged. There is no need for sterile, restrictive statutory and administrative provisions. The general powers of the central government over the state petroleum enterprise are sufficient.

Economic analysts will differ in their assessment of the success or failure of state petroleum enterprises of developing countries. The terms success and failure as used in relation to state petroleum enterprises must be understood in a special sense. Had some of these enterprises been owned by private persons and not the government, they would either have had to be dissolved or declared bankrupt. However, this type of failure hardly applies to a state owned enterprise. Further, the performance of a state owned enterprise has to be judged not only in terms of profits and losses, but also in terms of the wider socioeconomic and political objectives that may have been drawn up for it. This is especially true in terms of their role as instruments for stimulating the economic growth and development of the country.

But in many cases, the state petroleum enterprise has been so poorly endowed that it succeeds neither in terms of the conventional profit-earning criterion, nor in terms of generating employment, stimulating new activities, or assisting the overall economic development. Some are stillborn. Others survive as anemic and undernourished offspring languishing in a state of semiactivity. But such a result is neither desirable nor necessary for a state petroleum enterprise. One of their chief safeguards is the adequacy of the financial provisions, and these should be judged in terms of the tasks required of the state petroleum enterprise and the political and economic independence that the country wishes to maintain.

This article has deliberately avoided the temptation to indicate a minimum level of equity or provisions for reserves, and draws attention only to the fact that costs of exploration, drilling, production, refineries, and related investments have grown rapidly in recent years.

It is important that these be matched by correspondingly increased equity and other financial provisions, and by earnings and provisions for debt servicing.

It may be argued that financial inadequacy can be tolerated, even offset, by recourse to production sharing and other agreements with the transnational or independent corporations that carry the risks of exploration and, in some cases, all of the capital costs of development. There is, indeed, room for this type of cooperation and joint venture arrangement between state petroleum enterprises and transnationals. But transnationals are not philanthropic organizations. They are in search of profits, power, and influence. Their attitude towards a strong, well-endowed state petroleum enterprise will be different from that towards a weak and vulnerable company that can employ no leverage and is visibly dependent on external sources for finance and expertise. A financially strong state petroleum enterprise is, therefore, essential for maintaining its own corporate integrity, and its economic contribution to the nation, and for the substance of national sovereignty for the developing country itself.

This paper will have served its purpose if it succeds in stimulating interest in a topic that lies at the root of a problem that is of paramount importance to all the countries of the developing world - be they large integrated producers, refiners, or small marketers. Economists and others interested in this field should undertake further investigations into and analysis of this matter, and governments of the developing countries should begin the internal examination of this matter. Ultimately, the topic treated in this paper should become the subject of an objective discussion at an international forum, with all the developing countries and their state petroleum enterprises making contributions.

NOTES

(1) "Evolution of Nigerian National Petroleum Corporation," Nigerian National Petroleum Corp., Presented at OPEC Seminar on the Present and Future Role of the National Oil Companies, Vienna, October 10-12, 1977.

(2) "New Statutes for Oil Sector Companies approved, " Middle East Economic Survey, October 24, 1977.

(3) Harry O'Conner, World Crisis in Oil (London: ELEC Books, 1963), p. 189.

(4) Simon G. Hanson, "The End of the Good-Partner Policy," Inter-American Economic Affairs 14, no. 1 (Summer 1960): 79 et seq.

(5) Michael Tanzer, The Political Economy of International Oil and the Underdeveloped Countries (New York: Beacon Press, 1969), p. 355.

(6) W.J. Levy, "Basic Considerations for Oil Policies in Developing Countries," Techniques of Petroleum Development, Proceedings of the United Nations Interregional Seminar on Techniques of Petroleum Development, New York January 23-February 21, 1962 (Sales No. 64. II. B.2.), p. 328.

(7) See "The Role of Foreign Governments in the Energy Industries," Office of International Affairs, Dept. of Energy, Washington, D.C., October 1977.

(8) Ibid., p. 363.

(9) Ibid., pp. 363, 365.

(10) Tanzer, The Political Economy of International Oil, p. 299.

(11) Robert Engler, The Brotherhood of Oil: Energy Policy and the Public Interest, (Chicago, Ill.: University of Chicago Press, 1977), p. 33.

(12) UNCTAD - Bulletin of Peace Proposals, 7, no. 3, (1976).

(13) Charles R. Dechert, Ente Nazionale Idrocarburi: Profile of a State Corporation (Leiden: E.J. Brill, 1963).

(14) Fortune, August 1977.

(15) Dechert, Ente Nazionale Idrocarburi.

(16) Ibid., App. 1, p. 97, Note.

(17) Ibid., App. 1, p. 94, Note 1.

(18) Ibid., p. 8.

(19) See: "CEPE (Ecuadorian State Oil Corporation)" presented at OPEC seminar on The Present and Future Role of the National Oil Companies, Vienna, October 10-12, 1977, pp. 12-13.

16 Energy Utilization for Economic Development in Developing Countries and the Role of the State Petroleum Enterprises

D.H.N. Alleyne

This chapter draws attention to two important questions as they affect the developing countries in the context of the New International Economic Order. The first concerns energy consumption in developing countries and how it can be systematically employed for the promotion of economic growth and development. The second concerns the role of the state petroleum enterprise not only in the traditional operations of finding and producing, refining and marketing petroleum for revenue purposes, but especially in the promotion of economic growth and development by stimulating the utilization of energy resources directly within the economy, and by fostering the establishment of other industries through a process of backward and forward linkages with the petroleum industry as the growth point. (1)

The two questions are very closely linked in this examination and, though at first one attempts to treat just one and then the other, it will become necessary to marry them quite early in the analysis.

It is clear from an examination of various studies and opinions on the question that energy consumption by developing countries is not generally considered as a matter of great importance. The report of the Workshop on Alternative Energy Strategies (WAES), entitled ENERGY: Global Prospects 1985-2000, is concerned mostly with the energy supply/demand picture for the industrialized world. (2) While the report treats the demand for energy in the developing countries at certain points within the body of the publication, it is in Appendix I, "Energy and Economic Growth Prospects for the Developing Countries: 1960-2000," that this question receives particular attention.

The report points to the difficulty in obtaining data on energy consumption, especially in those countries where nonconventional fuels are used (as in India). It, therefore, concentrates on commercial energy and points out that: "During the period 1960-1972 the developing nations more than doubled their consumption of commercial energy and increased their demand for electric power by 25 percent." Forecasting of future energy demand is done in the following manner:

Energy demands have been estimated by first projecting economic
growth rates for the developing countries, consistent with the WAES
scenario cases.... once economic growth rates are produced the
energy required to sustain those levels of economic growth are
estimated.   This was done by first examining the historical
relationship between regional economic growth and energy consumption. (3)

Three observations need to be made on this approach: 1) It is based on
economic growth rates as the prime mover in the process with energy
demand as the result. In fact, however, energy demand or utilization
plays a significant role in determining economic growth. 2) It deals
with economic growth and does not mention economic development,
which is growth plus social and economic change. This leads to the
observation that 3) it is too closely confined to "the historical
relationship between regional economical growth and energy consumption." It is this "historical relationship" that the New International
Economic Order is seeking to change in so many spheres. But it is not
only among the people of the developed world that historical roles and
relationships are held to be virtually unchangeable.
    The following statement appears, unintentionally, not only to
reinforce the argument for the international division of labor to the
detriment of the developing world, but places the seal of Divine
Ordinance upon it, much in contrast to the convictions of this author.

In his Wisdom, it appears that God has bestowed upon many poor and
undeveloped countries the wealth of tremendous reserves of petroleum, to provide a means through which they may raise the living
standards of their people to a higher level. It may well be, also, that
through the separation of the areas of utilization and possession, it
was God's intention to provide a means through which it would be
possible for mankind to achieve an atmosphere of mutual understanding, cooperation, and consideration of interests. (4)

Of course, one also has to judge the statement in the more specific
context in which it was made, the struggle for larger revenues and
amended terms from the international operating comapnies, the
relatively low level of economic activity, and the absence of the
infrastructure without which the next steps to economic growth and
optimum utilization of natural resources become difficult.
    However, whether in any given country the appropriate timing for
the shift of emphasis is now or some years hence, the important matter
for consideration is the question of the "separation of the areas of
utilization and possession of oil and gas." If such a thesis were to be
accepted, the developing countries, especially those possessing resources of petroleum and natural gas, could at once resign themselves to a
position of perpetual economic inferiority. Recent estimates show the
extent to which the energy requirements of the developed world have
been increasing and, within this context, the increasing role assigned to

petroleum and natural gas. A comparison between the energy consumption of the developed and the undeveloped countries further emphasizes the point:

> To most of the people of the world it has become irrevocably clear that they are not getting all the goods, services, and rights they might possess. They perceive a sizeable gap between a society of scarcity and of abundance, between what is and what might be. For underindustrialized regions to close this gap, cheap and plentiful high production energy is necessary. The underdeveloped countries of Asia, where nearly one-half of the people of the world live, consume energy from coal, petroleum and waterpower at a rate equivalent to burning slightly over one hundred pounds of coal per person per year. This is about the amount consumed in the United States per person in two days. (5)

The question for examination, therefore, is the usefulness of substituting for the traditional pure income or revenue approach a process that uses the revenues from oil and gas as well as physical development through forward and backward linkages based on the petroleum sector.

On this question, it is interesting, once again, to refer to the WAES Energy publication. (6) While it recognizes that an approach based upon assumed economic growth rates as the main or sole determinant of energy demand has its limitations, the report is reluctant to introduce as a causal factor the deliberate promotion of energy utilization by the developing countries; nor does it concede that such stimulated energy utilization involving new energy-based or related industries will lead to accelerated economic growth and economic development by bringing about changes in the structure of the economy and of the society. Note, for example, the following extract from this report:

> Obviously using such a simple relationship between economic growth and energy growth is inadequate in some respects, in that it fails to include factors that may significantly affect energy consumption - such as changes in industrial structure, or increased mechanization in agriculture. Alan Strout, for instance, considers the production of a small group of key energy-intensive materials (such as iron and steel, cement, aluminum, etc.) which, when combined with using energy weights, are an indication or measure of the "energy intensiveness" of a country's industry. For developing countries achieving rapid industrialization, the production of energy-intensive goods would have to grow even faster than normal, and income elasticities would accordingly be higher. (7)

This raises the important issue of whether developing countries will choose to develop energy-intensive industries. The WAES publication examines the suggestion that a significant transfer of industry to the developing countries endowed with energy resources may occur. The examples given are in relation to the transfer to Brazil of aluminum smelters to utilize that country's hydroelectric resources and the export of chemical plants to energy-rich OPEC countries. On this question, however, the report continues

> It is the assumption of this paper that while such transfers may occur, there will be no major transfer of energy-intensive industries from the developed to the developing world. Furthermore, the immense complications of such a transfer, both for the developed and the developing, suggests that most developing countries will continue to be net importers of these products. (8)

What this part of the WAES Energy Report really boils down to is the promulgation in another form of the argument for maintaining the status quo and the traditional relationships. In particular, it implies support for the thesis about "the separation of the areas of utilization and possession" of petroleum resources.

As against that position, this chapter asserts that it should be the objective of all oil-producing developing countries to convert themselves into producers and consumers of energy, and participate in the utilization of their energy and raw material sources. Further, this transformation should be assisted consciously and deliberately by the revenues derived from the initial stage when the country is merely a producer and exporter of oil and gas resources in their crude form. For the nonproducers of petroleum, the exercise is more difficult but is, nevertheless, essential.

Recent calculations have indicated the growing demand for energy resources in the developed countries of the world. In fact, degrees of development can be meaningfully indicated not merely in terms of national income figures (which are an end product of economic activity), but also in terms of the kinds of economic activity pursued by countries, as well as in terms of the per capita consumption of energy.

> Finally, we shall have to consider whether it is sufficient to measure the degree of "backwardness" or "advancement" of different groups of people merely in terms of the relative distribution of final incomes among them; or whether the pattern of distribution of economic activity among the different groups and the different roles they play in economic life might not in the long run offer a more significant clue to the future potential development of each group. (9)

In this exercise of gearing the economy to utilizing hydrocarbons, time is of the essence. It must not be concluded, however, that there is necessarily a simple, direct correlation between high energy consumption and a high standard of living. Certain European countries such as Denmark, Sweden, and the Federal Republic of Germany have about the same GNP per capita as the United States, but consume about half the energy per capita. (10) Today and for the next three or four decades, the main energy source is, and will be, petroleum and natural gas. Thereafter, nuclear energy may tend to become increasingly important.

(A relatively high percentage of all new power stations under construction or being planned in the United States are nuclear. There is a similar trend in Europe. The problems of pollution and safety will, however, place serious obstacles in the way of this development.) This shift will take place if, all things considered, the environmental problems of nuclear power are resolved and it becomes more efficient in physical and economic terms. Should the developing countries fail to take advantage of the essential resources with which they have been endowed by nature, they would be neglecting an opportunity for economic advancement which is not likely to recur. If the sufficient supply of petroleum and natural gas resources is too long delayed, the producing countries may well find that when they are ready to come into the energy and raw material utilization phase of economic development, they are once again hopelessly outstripped by developed countries that have moved on to another energy source. The gap between them may even have widened. In such a situation, they may find also that the supply of an alternative and perhaps more efficient energy source may have reduced the prices of their petroleum products to levels that, for countries dependent entirely on the pure revenue approach to petroleum, would be catastrophic.

The impression should not be conveyed, however, that gearing the economies of the developing countries to a greater utilization of petroleum resources for the promotion of economic growth and development is an easy or simple task, or one that can be achieved at great speed in most of those countries. Such a step contains within itself formidable obstacles. Every new industry or project will have its planning, financial execution, and operational teething problems. However, the existence of certain facilities even prior to the planning and execution of these projects will greatly enhance their prospects' success.

Among the more important of these infrastructure requirements are the following:

1. Public utility facilities:

   a) Water supply - most industries require the ready availability of large quantities of water, even where the minimum acceptable degree of purity may vary from industry to industry. Arthur Lewis stresses the importance of water in the promotion of economic development and concludes: "One simple test of the quality of a development plan is to see what it says about water." (11)

   b) Internal communications - to facilitate prompt decisions and the transmission of information for the transport of raw materials and finished goods.

   c) Electricity - energy in a most flexible form.

   d) Port facilities - capital goods, most of them of great weight and size initially will have to be imported. Adequate port facilities would assist the process.

2. Educated people, at the level of the general population and in terms of the professional, scientific, technical, and skilled artisan personnel that would be required for energy-based and related industries.

The development of an industrial base cannot be divorced from the overall socioeconomic development of the country. On the specific question of basic education, a recent sample of developing countries shows literacy rates ranging from 7 to 95 percent of the population over 15 years of age. In about 50 percent of such countries, only 20 to 40 percent of the population was classified as literate.

But the absence of ideal conditions does not mean that no start can be made. Special landing ships and barges can, to some extent, overcome the absence of ports; in most countries water wells can be substitutes (temporary or permanent) for large central water collection and treatment plants; modern electronic communication systems can be installed for limited but effective area and distance coverage. Some countries that are represented as having rather low literacy rates (for example 20 percent) have, in fact, been able to start massive industrialization programs and have developed a corps of highly trained personnel at all levels to plan, execute, and operate their new industries.

It must be clear from the foregoing, however, that not all countries are equally able to make the utilization of oil and gas the basis for promoting economic growth and development. It is possible to make a very rough and arbitrary classification of countries in descending order of capacity for achieving this objective. First, there are two broad classifications: the oil-producing developing countries, and the non-oil producing developing countries. Under the first, one can then have at least two lists:

1. The large oil and gas producers with physical infrastructure and with technically and professionally trained personnel; with large physical area, a large population, and a potentially large domestic market that can absorb some of the products from these industries. The large petroleum production permits allocation of petroleum resources for domestic requirements leaving a surplus for exports in the form of crude oil, gas, and refined products to earn revenue and, even more important, foreign exchange for financing the general infrastructure development and specific projects. Many combinations of levels of production, land area, and population size will be possible. The one prerequisite for inclusion in this category is large production in excess of domestic requirements (broadly defined) to permit the earning of substantial revenue and foreign exchange.

2. The medium to small producers with some or all of the characteristics mentioned above, but on a more modest scale. In fact, a medium to small producer having a small physical area, but with essential infrastructure, a highly literate population,

and a corps of qualified personnel may very well make the industrial step more easily and quickly than some of the large producers.

In the second category, one can also have two lists:

1. Countries with large area and population, essential infrastructure developed over a number of years, nonpetroleum industry, an educated population, and a generally viable economy, even though per capita income may be too low to remove them from the class of developing countries. The main problem of these countries will be the foreign exchange cost of imports of oil and gas. With judicious management, however, the task is not impossible, especially if selective barter or other arrangements can be made.

2. The small developing countries, some of them landlocked, with relatively small populations, low levels of literacy, etc.

These latter are really the countries that have the greatest problem and will need help from others more fortunate. But any help given need not be considered as a gift, It can, in fact, be seen as a long-term investment preparing the markets of these small countries to purchase some of the products that will flow from the industrialization of the larger oil-producing countries. Such arrangements obviously involve the governments of the developing countries, and their role in this matter of industrialization on the basis of oil and gas resources will form the greater part of the succeeding section.

In recent international developments, there are three topics of particular importance to our subject: 1) general recognition and acceptance of the principle of permanent sovereignty over natural resources, 2) the greater role of the governments of developing countries in petroleum affairs, and 3) the rapid creation of state petroleum enterprises. The latter have become instruments through which governments of developing countries can formulate and implement their policies and plans for owning, controlling, developing, and managing petroleum resources and making them contribute to the fiscal and physical development of the countries.

With regard to the question of permanent sovereignty, the various resolutions of the GeneralAssembly of the United Nations are relevant. They start with Resolution 523 (XI) of January 12, 1952, which identified the developing countries' "right to determine freely the use of their natural resources," and the need for them to "utilize such resources in order to be in a better position to further the realization of their plans of economic development in accordance with their national interests." Subsequently, Resolution 1803 (XVII) of December 14, 1962 declared, inter alia, that "The free and beneficial exercise of the sovereignty of peoples and nations over their natural resources must be furthered by the mutual respect of States based on their sovereign equality."

With these and other resolutions on the general question, action by developing countries with respect to their petroleum resources that formerly attracted the anger and retaliation of the developed world - Mexico (1938) and Iran (1951) - became "respectable." Such governments could face the practical problems of managing their petroleum industries and economies without the harsher, more obvious, retaliatory action of the developed countries and their transnational corporations.

One of the first problems to which developing countries had to address their thinking was how to stop the waste of natural resources, especially natural gas, that was taking place under the transnational companies who recovered their costs and made large profits from the crude oil and the refined products alone. Such a situation was unacceptable in that this waste of resources went hand in hand with limited domestic utilization of oil and gas.

The earlier, traditional role of the government was to tax the export sector and use the proceeds for investing in other sectors. Experience has shown, however, that this approach to economic development has serious shortcomings unless it is supplemented by other governmental plans and strategies. Inflation soon undermines the value of the tax proceeds and there is often great misspending of these revenues. One alternative, therefore, is the possibility of making the petroleum industry the central growth point for the country's economic development by utilizing the energy and other physical characteristics of oil and gas together with the revenues from petroleum exports to develop other industries and promote other forms of economic activity. For this new process, the government requires a new instrument that would take the place of the foreign oil companies and also assume additional functions.

Given the new involvement of governments in the petroleum industry, the next question was what should be the nature of the instrument that those governments should employ to implement the new policies especially in the context of the New International Economic Order.

One alternative is to use a ministry or a government department. This approach may seem attractive to those who are primarily concerned with maintaining government control over the operations of the industry. There is always room for a ministry or a department of petroleum minerals. The departmental and ministerial organization is, however, geared to a special type of bureaucratic function and system. It has not shown itself capable of adaptation to the operation of an enterprise that must exist side by side and often compete with private corporations. The mental attitude of people employed in a civil service is different from that of people engaged in private enterprise. The principles, procedures, constraints, and objectives are different.

Many governments have recognized this and have, therefore, opted for the second alternative, the creation of national petroleum companies. Iraq, for example, in the statute establishing the Iraq National Oil Company, specifically provided that "the company shall not follow the financial and accounting principles applied in the government in accordance with other laws and regulations."

Various countries, for reasons best known to themselves, have not followed fully the strict corporate form and principle. It has, therefore, been necessary to coin a new name, state petroleum enterprise, to describe all those enterprises or agencies established by governments to own, manage, and administer their petroleum resources and affairs. While one should note and respect the reasons for which governments may wish to depart from the strict corporate form, it is necessary, to point out the advantage of that form.

The corporate form has great inherent flexibility, permitting the entity to adjust promptly to rapidly changing situations in the international petroleum industry. It promotes a spirit of enterprise so that the corporation can examine and undertake new projects, and as a corollary to this, it demands prompt decision making. If there is one important drawback in most governmental systems, it is the length of time needed to make decisions on the most important and urgent matters. Additionally, the corporate form breeds efficiency. Strictly speaking, a national petroleum corporation should be subject to most, if not all, of the normal tests for efficiency required in a privately-owned corporation, plus those that examine its role as a "good corporate citizen." This expression is more exacting and meaningful in this context than is implied when used by a transnational corporation operating in a developing country. The national petroleum corporation or the state petroleum enterprise must demonstrate its contribution to the economic growth of the country and to the promotion of energy-based and related industires. Finally the corporate structure has come to be recognized as an excellent instrument for capital accumulation.

The national petroleum corporation or the state petroleum enterprise marries these advantages with the authority that comes with ownership and ultimate control by the sovereign state. Dr. Reyes Heroles, a Director-General of PEMEX, was of the view that a national oil company should be a "useful social tool." As Bermudez, another Director-General saw it, PEMEX was not only required to survive, but it had to "pace the Mexican industrialization" and "the economic development of Mexico."

It is proposed in the following section to examine some examples of attempts by developing countries (Iran and Mexico particularly) to convert their economies into consumers of oil and gas, thereby promoting the establishment of new industries. At the same time, the roles played by their national petroleum companies in achieving this advance into economic growth and development will be examined.

As with most other old oil-producing countries, the early official approach or policy in Iran centered around fiscal benefits. All the changes in concessions wrung from the operating companies have been directed towards obtaining a larger share of the cake. Even the nationalization resolution passed by the Majlis on March 15, 1951 was directed ultimately at obtaining for the government a larger financial return from the exploitation of the national resources. In the two decades following the nationalization of its petroleum resources and the creation of the National Iranian Oil Company (NIOC), the government took measures to give the company full and absolute control over the industry by 1973. Again, initially, the accent was on increasing the government take from a barrel of oil. In the mid 1960s, however, Iran launched a massive industrialization plan geared specifically to the development of fertilizers, petrochemicals, and electricity generation - all based on the utilization of oil and natural gas.

It was in 1908 that electricity was first introduced into Iran; the first plan was built by the Anglo-Persian Oil Company. Teheran received the first plant to be built outside of the oilfields, a unit of 6 Mw installed in 1934. By 1940, electric generating capacity was still only 50 Mw for the whole country. Under a succession of development plans, installed generating capacity and transmission and distribution systems were rapidly expanded. By the third plan, 1964-68, installed capacity was 894 Mw. Under the fourth plan, 1968-72, the installed capacity under ministerial auspices had risen to 2,094 Mw (and the country total, including installations by municipalities, and others was 3,335 Mw). Under the fifth plan, the target was to increase central government installed capacity from 2,094 Mw to 7,500 Mw. The policy included the installation of a national grid linking the various generating centers and transmission and distribution lines to give greater security and balance to the system. It aimed at expanding the electricity industry to meet industrial, agricultural, commercial, urban, and household needs. It also stressed rural electrification. In a link with the natural gas policy, which is treated next, the rural electrification plan provided for the installation of gas turbines for the villages to generate electricity, making feasible major movements in agricultural production, health, and sanitation directly affecting the overall stand-ard of living. It helped centralize small and scattered villages into larger units where possible, and to facilitate the functions of other government organizations in providing necessary public services in these larger villages.

Earlier in this chapter, reference was made to the waste of petroleum resources at the same time as there existed a tremendous need for the utilization of those resources for the development of the country. Iran was no exception. Up until the early 1960s, almost all the associated gas in the Khuzistan oilfields, run by the consortium, was flared. Between the start of the oil industry in Iran and 1971, some 2×10^{11} cubic meters of natural gas have been flared.

What has the government and the NIOC done about that situation? After the nationalization and creation of the NIOC in 1951 came the settlement by the creation of the consortium in 1954. The year 1954 saw the first use of gas by the consortium through the transmission of 1.5 million cubic feet (mcf) per day from the Agha Jari oilfield to Abadan for use in the refinery. LPG was introduced in 1955. It was not until 1963, however, that the first pipeline was installed to transport gas from the flare stacks of Khuzistan to more distant urban centers, for example from Gach Saran to Shiraz.

In the initial stages, the gas utilization was confined to the southern part of the country near the oilfields. In many cases, an external stimulus is required to provide the domestic or internal advance. The external stimulus, providing gas to the northern parts of the country, was the 12 year natural gas agreement with the USSR. Previous to this, the capital and operating costs of constructing a gas pipeline to Tehran and the north of the country were such that the project was not undertaken. With this agreement and the large volumes involved, it was possible to plan a line that would serve both export and domestic requirements. This line, the Iranian Gas Trunkline (IGAT) stretching from Bid Boland to Astara, was completed in 1970 and carried some 1.9 x 10^{10} cubic meters of associated gas per year from the Khuzistan fields to the Soviet Union and to several cities in Iran as well.

The agreement with the Soviet Union provided for a supply of 6 x 10^9 cubic meters in the first year rising to 1 x 10^{10} cubic meters by the fifth year, so that there is a considerable quantity available for domestic Iranian requirements. Further, the very presence of such a line with built-in arrangements for introducing spurs from the main line means that actual or prospective economic development over a large area on either side of the IGAT can be assured of a supply of this important energy source. It will be interesting to see over the next one or two decades the extent of the transformation in the economy that will result from this project alone. Among the projects now served, or likely to be served, with gas by the National Iranian Gas Company (NIGC) are a million ton direct reduction steel-mill at Bandar Abbas and the petrochemical industry.

The NIGC has undertaken other objectives and projects as well. It aims to substitute gas for oil as a fuel in Iran to the maximum possible extent. And very well it might, for even with the gas utilization projects already undertaken, still more than 45 percent of the associated gas produced in Iran is still being flared.

Petrochemicals is one of the most capital-intensive industries. If a venture is successful, however, the value added in the various processes is perhaps greater than that achieved in most of the other sectors of the petroleum industry. Iran has decided that, as far as possible, it will use its hydrocarbon resources less and less as fuel and more as feedstock for petrochemical production.

A recent NIOC publication, Petroleum Industry in Iran, states the objective and rationale of the petrochemical industry:

> Currently, however, less than 5 percent of the world's hydrocarbons are used as petrochemical feedstocks. If Iran can succeed in diverting up to 10 percent of its total crude oil and gas production into petrochemical intermediates and final products, the value would be more than twice as high as the income derived at current prices from the sale of the remaining 90 percent of crude oil and gas. (12)

But Iran's move into petrochemicals started in 1961 with the installation, near Shiraz, of a plant for the manufacture of nitrogenous fertilizers. The real push into petrochemicals began, however, in 1965, when the National Petrochemical Company was established as a wholly owned subsidiary of NIOC. The company was given strict instructions and full powers and resources to develop a petrochemical base in Iran. Petroleum Industry in Iran refers to the petrochemical industry as "a strategic sector in the industrial development of Iran," one which plans to become a major world producer of basic petrochemical intermediate products based on the comparative advantage which it claims Iran enjoys in this field.

Other oil-producing countries may not be as well endowed as Iran with the factors of petrochemical production. Iran's successes in this field should, however, stimulate others to examine their own potential for and the advisability of entering one or other of the branches of this industry. Not all countries can or should engage in this activity. If they all did, many would fail. They should note, however, the factors cited by Iran as facilitating her own entry into the industry: "Abundant raw materials, sufficient investment capital, skilled managerial and technical cadres, a sizeable, rapidly expanding domestic industrial market, and access to advanced production technology through joint-venture agreements, coupled with assured export markets."

It is not proposed in this paper to provide a comprehensive list of the petrochemical projects already established and of those proposed for early implementation. Such a list would, indeed, be imposing. A description of the Iran Fertilizer Company will suffice, however, to give an indication of the scope of the country's petrochemical industry.

The first fertilizer plant in Iran was started near Shiraz in the early 1960s. Table 16.1 shows present production data, including new products started in 1975, and indicates what will be the additional products after completion of new plant expansion in 1978.

TABLE 16.1. Petrochemical Production in Iran

(millions of tons per year)

	As of 1975	Additional products after 1978 expansion
Ammonia	-	400,000
Urea	52,000	500,000
Ammonium Nitrate (26 percent)	30,000	250,000
Nitric Acid (100 percent)	4,000	200,000
Soda Ash	63,000	-
NPX mixed fertilizer (1975)	50,000	-
Sodium Tripolyphosphate (1975)	30,000	-

What is interesting here for a study of the role of petroleum and of the state petroleum enterprise in the development process is not merely the expansion of output, but the implication of this increase in output for the agricultural sector and, thus, for the rest of the economy. All the products of this complex are consumed and will continue to be consumed by the domestic market. Given Iran's land mass (not all of it arable, of course) and the size of its population, there may very well be need for further expansion of production through the establishment of new fertilizer plants located to serve other areas of the country more directly. But the resources are available to meet future demand, which was expected to increase from 4.5×10^{11} tons per year in 1974 to about 1×10^{12} tons per year in 1978.

The new emphasis on fertilizers is an interesting development. As has been observed,

> Further, a recent publication by David Missess, "Iran: Oil at the Service of the Nation" contains no reference to a positive policy of integrating the oil and gas industries, through the linkage procedures, with the rest of the economy. For example, in discussing measures being undertaken "to make agriculture more productive and efficient" he mentions that "the use of water resources for farming purposes, irrigation and drinking water for livestock, is to increase by 15 percent over the Plan period." There is no mention of fertilizers. This may have been only an oversight but it does indicate that even in one of the most progressive petroleum producing countries and at the level of the writers of petroleum policy the benefits from oil through linkages may result from an accidental rather than from a deliberate and planned approach. It may be argued that once the benefit is achieved it is immaterial that its advent is accidental. The point, however, is that it is more likely to be achieved if it is consciously planned and pursued. (13)

In fact, it is clear from the more recent growth in the production of fertilizer and the plan for the future that the initial steps in the production and utilization of this very important input into agricultural and national development grossly underestimated current and prospective national requirements. It is now clear that its full growth and role were not at first foreseen or envisaged.

A comparative study of the Iranian and the Mexican approaches to the use of petroleum and the role of their respective state petroleum enterprises in promoting economic development will, perhaps, one day be presented in a more comprehensive manner than is possible in this short chapter. However, some brief impression will be given here.

It is first necessary to make allowances for the fact that these two countries started to utilize petroleum to industrialize at quite different times. Iran started in the 1950s with the nationalization of the petroleum industry and the establishment of the National Iranian Oil Company (NIOC) in 1951 and the creation in 1954 of a consortium to break the impasse resulting from the nationalization. This was at least 12 to 15 years after Mexico had already started to establish the base for its take-off into self-sustained economic growth.

In both countries, the respective state petroleum enterprises were given considerable powers and scope for action. The financial provisions for NIOC were, however, more generous than those for PEMEX. The latter was not only kept short of funds but was required to promote and subsidize prices for certain products. PEMEX was also required to make substantial financial contributions to the central government. Whereas Iran was soon in the league of the big petroleum producers, with large surpluses for export, Mexico was struggling for domestic self-sufficiency. While the emphasis in Iran was to increase the value of export and the government take, the emphasis in Mexico was to achieve greater domestic utilization of oil and gas for promoting economic development and diversification.

Both countries faced foreign opposition to their nationalization policies. In the case of Iran, the Anglo-Iranian Company took steps to impose a shipping boycott of Iranian oil, to block certain Iranian financial assets abroad, to embargo essential parts and equipment needed by Iranian oil facilities, and to take legal action against possible independent buyers of Iranian oil. Similar pressures were imposed on Mexico and PEMEX. The Standard Oil Company of New Jersey had predicted that "Mexico cannot operate the industry alone." Further, the capital markets of the world frowned on the confiscation and threatened dire consequences regarding the future flow of capital to Mexico.

What is important here is an assessment of the actions and approaches, sometimes similar but often quite different, taken by these two countries in the face of this kind of opposition. Both wished to use oil and gas for economic development. Both set out to use state petroleum enterprises as the instrument for achieving this objective. For a time, there the similarity ended.

Iran stressed the revenue approach - depending upon measures to break the blockade, so to speak, and earn revenues and foreign exchange from exports. The nature of Iran's social and economic base may have made this course inevitable. In the critical period of 1951 to 1954 the national policy aimed, inter alia, at serving relatively modest domestic petroleum needs with the help of national technicians and native management; reworking the development plan, within the confines of a self-reliant oilless economy, in such a way that an austerity budget could sustain the government on existing revenues; ear-marking potential oil incomes to be used only for increased capital formation in the future; and fighting for an independent inroad into the world market. (14) The emphasis on oil revenues predominated until well into the 1960s.

In the case of Mexico, "nationalization of oil was used to spur the strengthening of poverty-ridden sectors of the economy and to create a long-neglected domestic market in place of the previous corporate emphasis on crude oil exports." (15)

It is interesting to note the policies and objectives of two Directors - General of PEMEX, A.J. Bermudez and Reyes Heroles. In opposition to the view expressed by one of the major oil companies after nationalization, Bermudez set out to prove that Mexicans could operate the industry, and he succeeded. Further, PEMEX had to pace the Mexican industrialization and the economic development of Mexico. In fact, it did during the period when Mexico's economic advance was more rapid than that of most of the western world. Bermudez quickly realized that the requirements of PEMEX could serve as the take-off point for creation of a whole cluster of industries around it. By 1957, Mexican industry was supplying half of the requirements of the oil industry, compared with the import of virtually all its requirements in 1947. (16) PEMEX and the oil industry successfully fostered investment in the iron and steel industry, pipeline manufacture, fertilizers, and petrochemicals.

Heroles saw a state petroleum enterprise as a useful social tool. He said, "There are some areas of the country where there is no justification, from a profitability point of view, for a gas line, however, its laying breaks the vicious cycle hampering the progress of those areas." Iran postponed serving the northern areas of the country until the gas agreement with the USSR provided for the building of the Iranian Gas Trunk Line (IGAT) in 1970, thereby making it economically feasible to serve those areas with natural gas.

One of the essential points to be noted in attempting a comparison between the Iranian and the Mexican experiences is that there is much more published data and in-depth analysis of Mexico than of Iran. In addition to the contributions noted above, one can find a large number of works dealing with various aspects of the Mexican economy. For example William P. Glade and Charles W. Anderson found that in Mexico "the petroleum industry has been employed as a multifaceted instrument of development" contributing "to a wider diffusion of technical and managerial skills among the domestic labour force and provided opportunity/incentives for advancement to more productive types of labour." (17) Tanzer wrote that "petroleum has been a (if not the) leading sector in the Mexican development process." (18)

As indicated in an earlier section, Iran launched its own national petrochemical company in 1964 and has since made great strides in that industry. Also in 1964, a United Nations Survey reported that 40 percent of the petrochemical capacity in operation or under construction in Latin America was owned by PEMEX.

In 1968, it was reported that "The official policy of deliberately stimulating the demand for natural gas has contributed in no small measure to Mexico's rapid industrial development, not least in the sphere of petrochemicals." (19) Mexico, with the second largest proven reserves of natural gas in Latin America (after Venezuela), was able, through PEMEX, to overcome the difficulties of distance and terrain and to utilize a high proportion of its natural gas. Out of a total production of approximately 1600 mcf per day, only the relatively small amount of 360 mcf per day was wasted.

The recent large discoveries of oil and gas by PEMEX not only ensure self sufficiency for a long time to come, but give the country the capacity and the opportunity for large-scale exports of both commodities. Mexico, therefore, like Iran, can begin to earn a large amount of revenue and foreign exchange from her petroleum resources. Like Iran,

Mexico can now pipe large quantities of natural gas to an industrialized neighbor, in this case the United States. One can only hope that this new source of wealth does not undermine the earlier policy and resolve under which Mexico, to a large extent, had succeeded in 'internalizing' the dynamics of its economy and no longer depended primarily upon external stimuli.

While Iran and Mexico have been selected for special study here, it must not be overlooked that other developing countries also have been making similar efforts - some with marked success. (20) On the one hand can be seen the industrializing effort of a large oil producer and exporter like Venezuela and on the other hand there is the case of Brazil, a net importer of petroleum which, in spite of this disability, appears to be the prospective industrial giant of Latin America. Again, there is Trinidad and Tobago, very old petroleum producers. Although small in physical size and in output of petroleum, Trinidad and Tobago are on the threshold of rapid economic development. This will only be achieved, however, by careful planning and efficient execution of the task of establishing the proposed energy-based industries.

Africa south of the Sahara requires particular attention. A recent paper pointed out that, of all the regions of the world, that part of Africa below the Sahara demonstrated the slowest growth in the development of refining capacity. (21) The region contains a large number of the least developed and land-locked nations whose poverty is reinforced by lack of ready physical access to supplies of energy. Even Nigeria, the largest crude oil producer in the area, with a population estimated at some 80 million people, had in 1975 a refinery capacity of only 60,000 b/d as compared to production of some 2 million b/d. There is, therefore, considerable scope for advancement both in Nigeria, which has the potential, and in the rest of the region where the need far outstrips the supply.

In Nigeria, plans have been drawn up and funds allocated ($7.5 billion for 1975-80) for substantial investment by the Nigerian National Petroleum Corporation in capital projects based on petroleum. It would be interesting to see what results would follow from careful implementation of a policy similar to Iran's or Mexico's but tailored, as it must be, to the peculiarities of the Nigerian situation. For example, what impact on the Nigerian economy and the society would result from the phased development of a pipeline system linking the supply centers of natural gas to the actual and potential areas of industrial development? What would be the result for food supplies and nutritional levels of large-scale production of nitrogenous fertilizers based on the large resources of natural gas now being flared, if as in Iran it is all consumed by the domestic agricultural sector or if some of it is made available to other countries in the region whose needs are also urgent? In short, what scope is there for cooperative ventures of this nature between the petroleum producers in Africa and the energy-short nations of that region?

The ability of the petroleum-producing nations to assist others will be circumscribed by their own success in promoting domestic economic development. It is important, therefore, that they gear themselves for

economic growth and development by joining the important club - the
group of countries that attract attention by either their total and per
capita consumption of energy or the annual rate of growth of
consumption of energy.

In most countries it will be a chicken and egg problem. However,
once the will exists, it is not too difficult to begin by a judicious
expansion of fertilizer production and of electrical generating capacity,
by fostering the growth of other industries that require either fuel or
energy in quantities that permit backward and forward linkages with the
petroleum sector.

NOTES

(1) W. Arthur Lewis, Development Planning: The Essentials of
Economic Policy, (New York: Harper & Row, 1966), p. 256.

(2) Workshop on Alternative Energy Strategies, Energy: Global
Prospects 1985 - 2000, (New York: McGraw Hill, 1977).

(3) Ibid., pp. 95-96.

(4) Ibid., p. 3.

(5) R. Engler, Politics of Oil: A Study of Private Power and
Democratic Directions, (New York, Macmillan, 1961), p. 5.

(6) Energy, p. 279.

(7) Ibid., p. 295.

(8) Ibid.

(9) Ibid.

(10) H. Myint, "An Interpretation of Economic Backwardness" Oxford
Economic Papers 6, no. 2, (June 1954).

(11) Lewis, Development Planning, p. 99.

(12) NIOC, Petroleum Industry in Iran, p. 20 .

(13) D. Alleyne, "Oil and Gas in the Development Process," Patterns of
Progress, (Trinidad and Tobago, Ten Years of Independence), 1972, p.
96.

(14) Jahangir Amuzegar, Energy Policies of the World: IRAN, (Newark,
Delaware: University of Delaware Press, 1975).

(15) Robert Engler, Brotherhood of Oil: Energy Policy and the Public
Interest, (Chicago, Ill.: University of Chicago Press, 1977).

(16) John Hickey, "PEMEX: A Study in Public Policy," and Antonio J.
Bermudez, "Doce Anos al Servicio de la Industria Petrolera
Mexicana," Inter American Economic Affairs, 14, (Autumn 1960):
73.

(17) William P. Glade and Charles W. Anderson <u>Political Economy of Mexico</u> (London: Petroleum Press Bureau, Ltd.) p. 26.

(18) Michael Tanzer, <u>The Political Economy of International Oil and the Underdeveloped Countries</u> (Boston: Beacon Press, 1969), p. 298.

(19) Ibid., p. 27.

(20) <u>Petroleum Press Service</u>, (December 1962), p. 27.

(21) D.H.N. Alleyne, "Possibilities and Opportunities for Cooperation Among Developing Countries in Petroleum Refining," <u>Petroleum Cooperation Among Developing Countries</u> (United Nations publication, Sales No. E.77.II.A.3), p. 56.

Appendix I:
List of State
Enterprises and
Related Government
Authorities

AFGHANISTAN

Afghanistan Petroleum Exploration Department,
Kabul

ALGERIA

Societe Nationale de Transports
et de commercialisation des
Hydrocarbures (SONATRACH),
Algiers

ARGENTINA

Gas de Estado,
Buenos Aires; Yacimiento Petroliferos Fiscales,
Buenos Aires

BANGLADESH

Petro-Bangla Oil Agency,
Dacca

BRAZIL

Petroleo Brasileiro, S.A.,
Rio de Janeiro

BURMA

The Myanmma Oil Corporation,
Rangoon

CHILE

Empresa Nacional De Petroleo (ENAP),
Santiago

COLUMBIA

Empresa Colombiana De Petroleos,
Bogota

EDUADOR

State Oil Company (CEPE),
Quito

EGYPT

Egyptian General Petroleum Corporation,
Nsar City,
Cairo; Mist. Petroleum Company,
Cairo

ETHIOPIA

Ethiopian Petroleum S.C.,
Addis Ababa

INDIA

Oil and Natural Gas Commission,
Dehra Dun; Hydrocarbons India Ltd.,
New Delhi; Indian Oil Corporation, Ltd.,
New Delhi; Indian Petrochemicals Corporation Ltd.,
Baroda; Oil India, Ltd. (partly government-owned),
Duliajan

INDONESIA

Pertamina,
Jakarta

IRAN

National Iranian Gas Company,
Tehran; National Iranian Oil Company,
Tehran

IRAQ

Iraqi Company of Oil Operations,
Kirkuk; Iraq National Oil Company,
Baghdad

KHMER REPUBLIC

Petronas Oil and Gas Agency

KUWAIT

Kuwait National Petroleum Company (60% government-owned),
Kuwait; Kuwait Oil Corporation,
Kuwait

LIBYA

Libyan National Oil Corporation,
Tripoli

MALAYSIA

Petronas,
Kuala Lumpur

MEXICO

Petroleos Mexicanos (PEMEX),
Mexico

MOROCCO

Bureau de Recherches et de
Participations Minieres,
Rabat

NIGERIA

Nigerian National Oil Corporation,
Lagos

OMAN

Petroleum Development (OMAN),
Mina-al-Fahal

PAKISTAN

Oil and Gas Development Corporation,
Karachi

PERU

Petroleos Del Peru,
Lima

PHILIPPINES

Philippine National Oil Company,
Manila

QATAR

Qatar General Petroleum Organization
(QGPO)

SAUDI ARABIA

General Petroleum and Minerals Organization (PETROMIN),
Riyadh

SRI LANKA

Ceylon Petroleum Corporation,
Colombo

TAIWAN

Chinese Petroleum Corporation,
Taipei

TURKEY

Turkey Petrolleri A.O. Genel
Mudurlugu (Turkish Petroleum Corporation),
Ankara

UNITED ARAB EMIRATES

Abu Dhabi National Oil Company,
Abu Dhabi

VENEZUELA

Petroleos de Venezuela (PETROVEN),
Caracas

Appendix II: Research and Scientific Institutes in Developing Countries

Professional Scientific Institutes Directly Involved in Petroleum Research and/or Training	Professional Scientific Institutions Indirectly Involved in Studies on Petroleum

AFGHANISTAN

	Department of Mines and Geology Kabul

ARGENTINA

Argentinian Institute of Petroleum, Direccion General de Yacimientos Petroliferos Fiscales (YPF), Buenos Aires	Consejo Nacional de Investigaciones Cientificas Y Tienices, Buenos Aires; Institute of Industrial Design

BOLIVIA

Association de Ingenieros Y Geologos de Yacimientos Petroliferos, Fiscales Bolivianos, La Paz	Academia Nacional de Ciencias de Bolivia (National Academy of Sciences), Bolivia

BRAZIL

Instituto Brasileiro de Petroleo (Brazilian Institute of Petroleum), Rio de Janeiro	Instituto Historico e Geografico de Samta Catarina, Floriannopolis; Instituto Historico & Geografico Paraibano, Paraiba

CAMEROON

> Office de la Recherche Scienti-
> fique et Technique Outer-Mer
> Centre (ORSTOM), de Yaounde.

CHAD

> Office de la Recherche Scienti-
> fique Outer-Mer Centre (ORSTOM)
> Ndjamena.

CHILE

> Comite Nacional de Geografia
> Geodesica Y Geofisica, Santiago;
> National Commission for Scien-
> tific and Technological Research
> (CONICYT), National Institute of
> Technological Research.

CUBA

> Instituto Cubano de Recursos
> Minerales (IRCH), Havana; Cuban
> Institute for the Development of
> Chemical Industry (ICDM)

EGYPT

National Research Centre,   Geological Survey and Mining
Petroleum and Minerals   Authority, Cairo
(DOKKI), Cairo

EL SALVADOR

> Centro de Estudios e Investigaci-
> ones Geotecnicas, San Salvador

ETHIOPIA

> Geological Survey of Ethiopia,
> Addis Ababa; Geophysical Obser-
> vatory, Addis Ababa

GHANA

> Council for Scientific and Industri-
> al Research, Accra

GUINEA

Secretariat d' Etat de la Recherche
Scientifique, Conakry

HAITI

Conseil National des Recherches
Scientifiques, Port-au-Prince

INDIA

Institute of Petroleum Explora-
tion, Oil & Natural Gas Commis-
sion, Dehra Dun; Institute of
Reservoir Studies Oil & Natural
Gas Commission, Ahmadabad;
Institute of Drilling Technology
Oil & Natural Gas Commission,
Dehra Dun; India Institute of
Petroleum, Dehra Dun; R&D
Centre, Indian Oil Corporation,
New Delhi

Council of Scientific and Industrial
Research, New Delhi; Central Fuel
Research Institute, Dhanbad; Geological Survey of India, Calcutta;
National Geophysical Research Institute, Hyderabad; National Institute of Oceanography, Goa

INDONESIA

Indonesian Petroleum Institute
Jakarta

National Institute of Geology and
Mining, Jakarta; Direktorat Geo-
logi, (Geological Survey of Indonesia)
Bauding; Jawatan Pertambangan
(Institute of Mining), Jakarta

IRAN

Iranian Petroleum Institute,
Tehran

IRAQ

Petroleum Research Institute
Baghdad

ISRAEL

Institute of Petroleum Research
and Geophysics, Helen

National Council for Research and
Development, Jerusalem

IVORY COAST

Direction de la Geologie et de la
Prospection Miniere, Abidjan; Mini-
stere de la Recherche Scientifique,

Abidjan; Societe Pour le Development Minier de la Cote d'Ivoire
(SODEMI), Abidjan

JAMAICA

Scientific Research Council

KENYA

East African Industrial Research
Organization, Nairobi; Mines and
Geological Department, Nairobi

DEMOCRATIC PEOPLES REPUBLIC OF KOREA

Academy of Sciences, Pyongyang;
Geology and Geography Research
Institute

REPUBLIC OF KOREA

National Academy of Sciences,
Seoul; Geological Survey of Korea,
Seoul; Basic Science Research In-
stitute, Seoul; Korean Scientific
and Technological Information
Centre

MALAWI

Geological Survey of Malawi

FEDERATION OF MALAYSIA

Geological Survey of Malaysia,
Ipoh, Perak; Geological Survey of
Malaysia, Sarawak and Sabia; Na-
tional Institute of Scientific and
Industrial Research (NISIR), Kuala
Lumpur

MALI

Instituto de Recherche Scientifique
de Mali, Bamako

MAURITANIA

Direction des Mines et de Indus-
trie, Nouakchott

MEXICO

Instituto Mexicano del Petroleo, Mexico

Instituto Nacional de la Investigacion Cientifica-INIC, Enrico
Martinez; Academic de la Investigacion Cientifica, Ciudad; Aca-
demia de la Investigacion, Cientifica, Apartado, Mexico; Consejo
Nacional de Ciencia Y Technolo-
gia (CONACYT) (National Council
for Science and Technology),
Mexico

MOROCCO

Division de la Geologie, Rabat;
Institut Scientifique Cherijien,
Rabat

NIGERIA

Geological Survey of Nigeria,
Kaduma South

PAKISTAN

Pakistan Council of Scientific and
Industrial Research, Karachi; Fuel
and Leather Research Centre,
Karachi

PANAMA

National Academy of Sciences

PARAGUAY

Instituto Nacional de Investigaci-
ones Cientificas (National Institute of Scientific Research)

PERU

Instituto Geofisico del Peru, Apdo;
Servicio de Geologia Minera, Apdo

REPUBLIC OF THE PHILIPPINES

Bureau of Mines, Manila; National
Research Council of the Philippines, Rizal

RHODESIA

Geological Survey of Rhodesia, Salisbury

SENEGAL

Bureau de Recherches Geologiques et Minieres, Dakar; Office de la Recherche Scientifique et Technique Outre-Mer Centre ORSTOM de Dakar, Hann

SIERRA LEONE

Geological Survey Division, Freetown

SRI LANKA

Ceylon Institute of Scientific and Industrial Research, Colombo

SUDAN

Industrial Research Institute, Khartoum; National Council for Research, Khartoum

SWAZILAND

Geological Survey and Mines Department, Mbabane

TANZANIA

Mineral Resources Division, Dodoma

THAILAND

Applied Scientific Research Corporation of Thailand, Bangkok; Department of Minerals, Bangkok

TURKEY

Maden Tetkik ve Arama Entitusii MTA, Ankara; Turkiye Bilimsel ve Tchkmik Dokiimantasyon Merkezi (TURDOK), Ar•:ara

URUGUAY

Consejo Nacional de la Investigacion Cientifica y Tecnica

UNITED ARAB REPUBLIC

Petroleum Research Institute   National Research Centre, Cairo
Cairo

VENEZUELA

Instituto Venezolano de Petroleo
Quimica, Caracas; Instituto
Venezolano de Petroquimica,
Caracas

ZAMBIA

Geological Survey of Zambia,
Lusaka; National Council for
Scientific Research, Lusaka

Appendix III:
Training Institutes

A. Africa

1. Algeria:

a) Algerian Petroleum Institute with centers at Dar El Beida, Bourmerdes, Es Senia.

b) National Hydrocarbon Institute.

c) Ecole Nationale Polytechnique, El Harrach, Algiers. (Chemical Engineering and Petrochemistry)

2. Egypt:

a) Suez Institute of Petroleum and Mining Engineering (5 year courses)

b) El M... Training Centre, Cairo.

c) Faculty of Engineering, Cairo University.

d) Faculty of Engineering, Al Azhar University.

e) National Research Centre, Cairo.

f) Egyptian Petroleum Research Institute, Cairo.

3. Kenya:

Kenya Polytechnic (Department of Engineering and Science)

4. Libya:

a) Intermediate Institute for Oil Technology, Tobruk.

b) Advanced Oil Technology Institute, Tobruk.

c) College of Oil and Minerals, Tripoli University.

5. Morocco:
   a) Mohammed V. University, Rabat.
   b) Ecole Mohammedia d'Ingenieurs.
   c) Cherifian Petroleum Institute.

6. Nigeria:
   a) College of Technology, Yaba Lagos
   b) University of Ibadan (Geology)
   c) University of Ife, Ile-Ife
   d) University of Lagos

7. Tanzania:
   Dar Es Salaam Technical College

B. Middle East:
   1. Iran:
      a) Iranian Petroleum Institute, Tehran
      b) University of Isfatan (Department of Geology)
      c) University of Tehran, Tehran
      d) Arya Mehr Industrial University, Tehran
      e) Abadan Institute of Technology
      f) Tehran Polytechnic

   2. Iraq:
      a) University of Baghdad (Department of Geology and Petroleum Engineering.
      b) Higher Technical Institute, Baghdad.
      c) The INOC Training Centre, Baghdad
      d) Petroleum Research Institute, Baghdad
      e) Central Laboratories, Basrah

   3. Saudi Arabia:
      a) University of Petroleum and Minerals, Dahran.
      b) Saudi Arabian Institute for Higher Education, Mecca.
      c) Higher Institute of Technology, Riyadh.

C. Asia:

1. India:

   More than 100 universities and institutes of technology
   provide the basic training up to B.E., M.Sc., and Ph.D. levels
   in different disciplines.   Indian School of Mines, Dhanbad
   specializes in Reservoir and Petroleum Engineering.

2. Indonesia:

   a) Indonesian Petroleum Institute.

   b) Bendung Technological Institute.

   c) University of Indonesia.

   d) Surabaja Institute of Technology.

3. Pakistan:

   a) Pakistan Council of Scientific and Industrial Research.
      (Central and Regional Laboratories - Fuels).

   b) University of the Punjab, Lahore.

   c) University of Sind.

   d) Rawalpindi Polytechnic Institute.

   e) West Pakistan University of Engineering and Technology.

4. Thailand:

   a) Asian Institute of Technology, Bangkok.

   b) Technological Research Institute.

   c) Chulalongkom University, Bangkok.

Sources: World of Learning, 1970-71, 21st Ed. (London: Europe
Publications, 1971) and Personal Information.

Index

About the
Contributors

ERVIN LASZLO -- Director, UNITAR/CEESTEM Project on the New International Economic Order.

JOEL KURTZMAN -- Coordinator, Project to Create a New International Economic Order.

D.H.N. ALLEYNE -- Permanent Secretary, Office of the Prime Minister, Trinidad and Tobago.

LUDWIG BAUER — President, OMV, AG, Austria.

PAUL H. FRANKEL -- Chairman, Petroleum Economics Ltd., London.

PIET HARVONO -- President Director, Pertamina; Council Member, ASCOPE.

C.A. HELLER — Petroleum Consultant, New York.

ALI JAIDAH — Secretary-General, Organization of the Petroleum Exporting Countries, Austria.

M. NEZAM-MAFI -- Head, Department of Research and International Problems, Legal Affairs, National Iranian Oil Co., Iran.

C. VANRELL PASTOR -- Engineer, Secretary-General, ARPEL.

ARTURO DEL CASTILLO PEREZ -- Deputy Director of Economic and Industrial Planning Studies, Instituto Mexicano del Petroleo, Mexico.

V.V. SASTRI -- Director, Research and Development Division, Institute of Petroleum Exploration, Oil and Natural Gas Commission, India.

SAMMI SHERIF -- Board Member and Technical Advisor, Irani National Oil Company, Iraq.

A.H. TAHER -- Governor, PETROMIN, Saudi Arabia.

HASAN S. ZAKARIYA -- Interregional Advisor, Petroleum Economics and Legislation, United Nations.